Why Many Scientists Believe in God

Myron K. Loss

Why Many Scientists Believe in God

Second edition - October 2020

ISBN 978-1-4951-3814-0

Printed in Colombia
Impreso en Colombia
Por: Editorial Buena Semilla

Published by Encouragement Ministries
3408 NE 23rd Ave.
Ocala, FL 34479
encmin@gmail.com
570-765-2352

Additional copies are available for $12.00 that includes shipping to US addresses. Write for costs for overseas shipping.

40% discount to everyone on purchases of five or more, and 50% discount to bookstores on purchases of two or more.

Other titles by Myron Loss are available from Encouragement Ministries. All prices include shipping to US addresses:

Culture Shock – Dealing with Stress in Cross-cultural Living – 1982 – 175 pages - $12.00

A time-proven standard for preparing missionaries and other workers for the transition from one culture to another, and help them to survive the heavy stress.

The Reward of the Righteous – 1995 – 71 pages – $8.00

A study of the eternal rewards that God is preparing for those who are in Christ.

Letters from Heaven – 2015 – 101 pages - $11.00

Anthony Moyer dies and goes to Heaven and finds a way to write letters to his niece and other friends back on Earth describing what is happening in Heaven.

The World is Up-side-down – 2015 – 336 pages - $14.00

This is a deep Biblical study of the contrasts between the values of the world and the values in the Kingdom of Heaven. This is designed to help the Christian grow in his understanding and commitment.

God is with Us – 2015 – 404 pages - $15.00

This book, with over 1800 Bible references, answers many important questions such as: Who is God? How does He relate to man? Where does faith come from? Does God answer prayer? Can He change His mind? Does He know all of the future? Is the future already determined? It shows where Augustin and Calvin got it wrong.

Why Many Scientists Believe in God – 2017 – 245 pages, 2nd Edition 2020 – $13.00

The author examines many recent scientific discoveries that point convincingly to the fact that our world did not just happen by chance. Someone designed and created our universe and all that is in it. If you are not sure if God exists, you need to read this book.

Jesus, my Substitute – 2018 – 55 pages – $6.00

The secret for victorious Christian living is found only in complete surrender to the control of Jesus and the Holy Spirit. When Jesus is living his life for you, love, joy, peace, and the other fruits of the Spirit are all produced as natural fruit.

Can you Lose your Salvation? – 2018 – 49 pages – $6.00

What does the Bible say about this topic that has confused so many Christians? From Jesus's own words, you will see that he promises to lose none of his sheep.

Coming soon: **What happens when you die?**

Acknowledgments

Several friends have made suggestions and corrections to this text. I would like to acknowledge one friend in particular who helped to curb many of my excesses, to straighten out some of my scientific ignorance, and to suggest better ways of saying things. Dr. David Coombs is a Bible translator and an anthropology and translation consultant who has spent his life translating the Scriptures, teaching other Bible translators how to study minority cultures, and then consulting with them as they translate God's Word into the minority languages of Peru. Together, these translators are giving the speakers of those languages a chance to really know who God is.

Dr Coombs must be a very wise man, because he heartily approved when his son married our daughter.

I think that he should have written this book instead of me because he knows a lot more about science. But, as I told him, throughout my life there were always people who could do things better than I could, but they were never around when I needed them, or they were too busy.

Thanks, Dave for being available and for helping me do a better job in telling people many of the wonderful reasons to believe in God.

Table of Contents

Introduction

I do not feel obliged to believe that the same God who has endowed us with sense, reason, and intellect has intended us to forgo their use. Galileo

This book is an attempt to solve the question of God's existence for the person who is uncertain. It won't convince the skeptics, but will show the seekers of truth some of the overwhelming evidence that shows us that he is here.

Whether or not God exists should not be a mystery. There is overwhelming evidence that tells us he is present (throughout the book, I use the term "he" generically out of convenience since God, the Father, is neither male nor female and does not have a physical body).

After looking at all of this evidence, you may choose not to believe in God. You have the freedom to close your eyes to his existence, but then you are denying the overwhelming evidence and you cease to be a scientist in search of truth.

Almost everyone loves solving a good mystery. We like picking up clues and guessing who did the deed while eliminating the suspects who couldn't have done it. Sometimes the writers of mystery novels give us sucker clues to make us suspect the wrong person.

Eventually, if we're persistent and we make it to the end of the story, we usually find out who the guilty one really is, and sometimes it is a surprise. We all like to congratulate ourselves for our intelligence when we solve the mystery early.

In many ways, life is a mystery. Ever since man has been able to think and reason, he has been asking himself these questions: 1) Why is there something instead of nothing? 2) Who am I? 3) How did I get here? 4) Where did the world come from? 5) Where am I going? And 6) what am I supposed to do? The answers people have come up with vary greatly and sometimes seem very strange to us, but unfortunately for some people the questions have gone unanswered.

Good detective work involves examining clues, usually at the scene of a crime, and coming up with logical deductions that lead to an understanding of what happened. It can be very hard work to analyze seemingly insignificant details, piece them together in some rational form, and eventually have a reasonable explanation for everything.

Television has made a killing off of murder mysteries and detective stories of all kinds, like “Murder She Wrote,” “Diagnosis Murder,” “Columbo,” “Poirot,” “Sherlock Holmes,” “Murdock Mysteries,” “Law and Order,” “Father Dowling,” “CSI,” “NCIS,” and a long list of others.

They are the most watched and longest running programs on the home screen. Agatha Christie, the queen of murder mysteries, and many other authors have made a very comfortable living writing mystery novels.

Often, some of the portrayed characters have been blind to some of the most important clues. They concentrate on unimportant things that are like red herrings that only lead them down a rabbit trail when they aren’t looking for rabbits. When that happens, it leads to false conclusions, and often in real life to false convictions.

One good example of missing the obvious is the experience of Sherlock Holmes. It seems that he and Dr Watson had gone camping. They pitched their tent under the stars. In the middle of the night, Holmes wakes his companion and says: "Watson, look up at the stars and tell me what you deduce.” Watson says: "I see millions of stars and maybe quite a few planets among them. It may be true that a few of the planets are quite like Earth and there might be life on them." Holmes interrupts him and says, "Watson, you fool! Somebody has stolen our tent!"

Like this funny illustration, people trying to solve the mystery of life have often overlooked the obvious. Even before Darwin promoted the Theory of Evolution in his *On the Origin of Species* in 1859, men had considered the possibility that the existence of life was not brought about through designed creation, but rather through accidental circumstances.

Especially since Darwin, these “naturalistic evolutionists” have attempted to explain changes in life forms through natural mechanisms, while often recognizing the extreme difficulty of explaining by chance the origin of life or of the universe in which it lives.

They also cannot give us any hope for a benevolent conclusion to the whole process of life. It is like handing us a book to read that is missing the first and last chapters. They also cannot give us any purpose for our existence. A mystery book lacking a beginning and an ending only leads to a futile exercise in trying to unravel the plot, and makes us scratch our heads as to why the book was ever written in the first place.

For a about a year, on our satellite TV system, programs that I recorded got interrupted. I didn't know this until we were watching them later and then, unexpectedly we got the message, *"portions of this program are missing due to poor satellite reception"*

Sometimes, big chunks were missing, and sometimes, there was so much missing that it just jumped to the very end of the program without ever telling us how the story ended. It was very frustrating, but not nearly as frustrating as listening to people trying to explain the existence of life without answering either the origin or the purpose for life.

The purpose of this book is to present many clues about the origin and workings of our world, and let you decide where the world came from. You don't have to be a scientist in order to understand, but you do have to be able to practice logic. Two plus two is still four in detective work, so use the rational side of your brain to arrive at reasonable conclusions.

Let's pretend we are detectives and we are invited to help solve a crime. I will give you the scientific clues, and you will need to draw your own conclusions. At the end, you should be able to give a reasonable explanation for who is responsible for the event.

In a good old-fashioned murder mystery, we often start with a corpse, a butler, a maid, and some inhabitants present in the house at the time. We must decide if the dead person took his own life, if it was an accident, or if there was a second agent (or more) responsible for the death.

If the death was intentional, we must find the murder weapon, the motives that anyone might have had, and whether or not each suspect had opportunity or an adequate alibi.

We then carefully eliminate the suspects who could not have done it. If we find clues of forced entry, and each of the occupants of the house has an alibi, we must look outside the house for our murderer. In this way we gather all of the evidence, arrange it in a logical pattern, and eventually draw our conclusions.

In trying to solve the mystery of life, we find an existing universe and a planet full of intricate life, and we must determine whether or not the elements present could have had the power and the opportunity to create themselves, or whether there was some outside influence that is greater than our world.

We will take a look at some of the "natural" laws at work in our world and try to analyze where they came from: 1) Did they just happen to exist? 2) Would these laws automatically be present in any world? Or 3) did someone design the universe and our world and make the specific laws that regulate them?

Some basic principles for our detective work –

Science ("*Scientia*" from Latin - meaning knowledge) is the human effort to discover the truths about the world in which we live. A scientific study examines all of the available evidence, puts forth a hypothesis to explain it, and then tests the hypothesis under controlled conditions. A true scientist can only discover that which is already there. Until practical, measurable evidence is found to confirm a scientist's hypothesis, it remains only a theory, or an educated guess.

For instance, Einstein's *Theory of Special Relativity* that says that time slows down near the speed of light, proposed in 1915, was only recently confirmed by hard evidence being found in outer space. After questioning and evaluating his theory, most scientists accepted it to be true by faith in spite of there being no solid scientific evidence to confirm it until recent years.

A concise definition of "Science" –

Knowledge about or study of the natural world based on facts learned through experiments and observation. - Merriam-Webster Dictionary

A Law of Science is a basic, unchanging principle of nature; an observed phenomenon that has been subjected to extensive measurements and experimentation and has repeatedly proven to be unchangeable throughout the known universe (e.g., the law of gravity, the laws of motion, the law of buoyancy).

We will look at the ideal conditions in our world that are significantly unchangeable if life is to continue to exist. For example, if the distance from the sun varied by 10%, or if there were no moon, or no planet like our Jupiter, or no electromagnetic field surrounding the earth, the complex life on our planet could no longer exist.

The scientific evidence tells us that life was begun by something or someone. Even in the Big Bang theory, there was a time when our universe did not exist, and then a time when its existence began. Someone or something had to create and form the infinitely dense ball and the "dynamite" that caused it to explode and throw itself out into ever expanding circles. And someone or something had to create the stars and then strike the match that lit the "fire" that has illumined the skies ever since.

It is one thing to admire the stars and try to analyze them: it is quite another to wisely interpret the obvious. To be a good detective we must be able to deduce the most basic truths like Sherlock Holmes and realize that "someone has stolen our tent." The obvious conclusion about the existence of our universe and of our own life is that a very intelligent and powerful someone made it all.

Let's look at some of the clues about our world that demonstrate this truth. With each one, ask yourself the question, "What are the chances of this happening by chance?" If it couldn't have happened on its own, then who is responsible? Who could have made a world like ours? What do these clues tell me about the Maker? What is he like?

PS: This book didn't just appear on my desk one morning. I assure you that it took me many days of thought, learning, writing, editing, formatting, revising and checking before it came to this form. Then, I sent it to the printer who meticulously laid it out, printed the pages and then carefully put them all together. This book is not an accidental occurrence, but a product of intentional design, yet it is not nearly as complicated as one drop of water from the vast ocean of our world.

Section I –

The big picture

To contemplate the vastness of space is mind-boggling; we are so small and seemingly insignificant compared to the size of the heavenly nebulae. At the opposite end of the spectrum, we cannot see the finely tuned intricacy of protons and neutrons or tiny one-celled creatures with our natural eyes. Every time we become able to look farther at the big things or deeper into the little things, we find more and more complexity and beauty.

On the big end, Hubble's orbiting space telescope has revealed amazing pictures of distant things. These immense phenomena were always there in the sky, but now we are getting to see much more of the beauty of the big picture.

On the small end of things, the Human Genome Project was finished in 2003, and it mapped out the blueprint from which a human being is made. All of a human's genes together are known as their "genome." Dr. Francis Collins, the director for this ground-breaking international project, says:

"I led a consortium of scientists to read out the 3.1 billion letters of the human genome, our own DNA instruction book. As a believer, I see DNA, the information molecule of all living things, as God's language, and the elegance and complexity of our own bodies and the rest of nature as a reflection of God's plan."

No matter in whatever direction we look, we discover more complexity, more precision, and more remarkable design. Today's scientists are awash in overwhelming new data that show that we live in an incredible world, so complex as to allow no illusions about it happening by chance.

1.

The illusion of matter

Now it seems to me the place to start is at the beginning.
Perry Mason

Concerning matter, we have been all wrong. What we have called matter is energy, whose vibration has been so lowered as to be perceptible to the senses. There is no matter."
Albert Einstein

"If quantum physics hasn't profoundly shocked you, you haven't understood it yet. Everything we call real is made up of things that cannot be regarded as real."
Niels Bohr

Most people in our world make a huge but erroneous assumption – we believe that matter is a solid substance. The truth is that a solid substance does not exist in our universe. When you sit at your kitchen table you do not place your food on a chunk of solid matter. You are placing your food on a chunk of energy. And the food is also a chunk of energy. And you are a chunk of energy.

All matter in our universe is an illusion of solidity created by spinning electrons and protons and other particles. Most of these sub-microscopic particles are only electrical charges, and if they could be compressed, your kitchen table would be smaller than a grain of sand. Between the electrical charges, there are very long stretches of empty space that compose 99.99999% of every atom.

Matter only seems to exist because energy exists. This energy has different forms that we perceive to be different substances, but in fact are just different configurations of the same charged particles. The reason we can touch or see anything is because we are also composed of energy. As our energy interacts with the other energy around us, we perceive its presence and its structure through our physical senses.

Scientists keep searching for particles other than electrons, protons and neutrons. Some years back, they discovered quarks and leptons. Recently, in Switzerland, they found evidence of a sub-atomic particle called the Higgs-boson that is often called the "God" particle.

All particles in our universe have motion. It is this motion and also the electrical charge and other characteristics of these particles that appear to give them physical properties. If this motion were to stop, then the universe would cease to exist. It is believed by many scientists that at a temperature of absolute zero (-273° C) all motion would come to a halt. If all cyclic motion stopped, the universe would cease to exist because every known thing in our universe depends on this motion.

Having studied the atom as a lifelong physicist, I am telling you that there is no matter as such! All matter arises and persists due to a force that causes the atomic particles to vibrate, holding them together in the tiniest of solar systems, the atom.

Max Planck

The neutron is a non-charged particle existing in the nucleus of an atom along with a proton that is positively charged. Neutrons do not repel or attract each other because they are neutral in charge. However, they do have weight because they are composed of a proton, an electron and a neutrino.

A neutron is 1.00137841870 times heavier than a proton, which is what allows it to decay into a proton, an electron and a neutrino—a process that determined the relative abundances of hydrogen and helium and gave us a universe dominated by hydrogen. If the neutron-to-proton mass ratio were not precisely as it is, atoms would not exist and neither would we.

Examples of such "fine-tuning" in our world abound. If you could make a small change in the charge of the electron, for instance, or change the strength of the force of gravity or the strong nuclear force just a tad, the universe would look very different if it existed at all, and it would almost certainly contain no living things.

Until the time of Einstein, mass and energy were considered to be two separate things. In the "Special Theory of Relativity" Einstein demonstrated that neither mass nor energy were conserved separately, but that they could be traded one for the other and that only the total "mass-energy" was conserved. The relationship between the mass and the energy is contained in what is probably the most famous equation in science.

E = m c 2

Where m is the mass, c is the speed of light, and E is the energy equivalent of the mass.

Because the speed of light squared is a very large number, a small amount of mass corresponds to a huge amount of energy. Thus, the conversion of mass to energy could account for the enormous energy output of the sun and other stars.

In 1919, Einstein correctly described the equivalence of mass and energy as "the most important upshot of the special theory of relativity," for this result lies at the core of modern physics. Particles of mass are nothing more than conglomerations of energy.

Electrons, protons, etc. can only be observed by observing their electrical charges or their effect on other particles. Mass is deceptive. It appears to exist as something palpable, visible, etc. but in reality, it is mostly space and only energy.

Our world was formed from energy and power. Did you know that if you could take all of the electrical energy out of your body, there would be nothing there? From where did all of the energy in our universe come? Could it have appeared out of nothing?

Who gave the power and organized it into ordered structures? Does evolution explain the origin of energy? Can chance explain the structure of an atom? Who is the policeman who enforces the laws of motion?

2.

Dark matter and dark energy

Darkness cannot drive out darkness: only light can do that. Hate cannot drive out hate: only love can do that.

Martin Luther King, Jr.

Until the early 1900s, scientists and philosophers believed that our universe was eternal, that it had no beginning and that it maintained itself in the same basic condition forever. Then came Albert Einstein and Edwin Hubble who bravely predicted that the universe was expanding and that it had a beginning. Their idea of a beginning of the universe was revolutionary and rejected at first.

In 1931, Arthur Eddington summed up the reaction of many astrophysicists when he wrote: "Philosophically, the notion of a beginning of the present order of nature is repugnant to me. ... I should like to find a genuine loophole."

The notion of a beginning of our universe gradually became more widely accepted, starting with Edwin Hubble's convincing demonstration that the whole universe was expanding outward.

The debate between the steady state theory and the Big Bang theory raged until the sixties, when the discovery of the background radiation in the universe strongly supported the belief that the universe had a beginning that started with a powerful explosion (a Big Bang).

In the early 1990s, one thing was believed fairly certain about the expansion of the universe; it might stop its expansion and recollapse, or it might always continue expanding but at a decreasing rate. Gravity was believed certain to slow the expansion as time went on. The universe is full of matter and the attractive force of gravity pulls all matter together like the opposite poles of a magnet.

Then came 1998 and the Hubble Space Telescope observations of very distant supernovae that showed that, a long time ago, the universe was actually expanding more slowly than it is today. So, the expansion of the universe has not been slowing due to gravity, as everyone thought; it has been accelerating. No one expected this or knew how to explain it. But something was causing it.

While scientists were scratching their heads about this new evidence, it was also found that galaxies in our universe seemed to be achieving an impossible feat. They were rotating with such speed that the gravity generated by their observable matter could not possibly hold them together; they should have torn themselves apart long ago. Therefore, scientists concluded, there must be some invisible matter or force that is causing this phenomenon. Scientists, for want of a better name, called it "dark matter."

The Milky Way is shaped like a disk that is about 100,000 light-years across. The stars in this disk all orbit the center of the galaxy. The law of gravity says that the stars that are closest to the center of the galaxy — which is also its center of mass — should move faster than those out on the galaxy's edge. Otherwise, the closest stars would be drawn into the center of the galaxy.

Yet when astronomers measured stars all across the galaxy, they found that they all orbit the center of the galaxy at about the same speed. This suggests that something outside the galaxy's disk is tugging at those stars: dark matter.

Dark matter is a hypothetical type of matter composing approximately 27% of the mass and energy in the universe. Dark matter is an invisible substance that can only be seen through the effects of its gravity. Only 4-5% of the universe is comprised of visible matter that is observable by light. Everything that we can see is like the tip of the cosmic iceberg — it accounts for only about five percent of the total mass and energy in the universe.

Everything ever observed with all of our instruments, i.e. all normal matter - adds up to less than 5% of the universe. Most of the universe apparently consists of nothing we can see or measure directly. About one-quarter of the universe consists of dark matter that releases no detectable energy, but that exerts a gravitational pull on all the visible matter in the universe.

Dark matter and also dark energy can only be observed by measuring their effects on the visible matter around them. We know how much dark energy there is because we know how it affects the universe's expansion. Other than that, it is a complete mystery. After visible matter and dark matter, the rest of the universe is made up of dark energy, baryonic matter, and neutrinos.

Dark energy makes up approximately 68% of the universe and appears to be distributed throughout the vacuum in space. It is distributed evenly throughout the universe, and its effect is not diluted as the universe expands. The even distribution means that dark energy does not have any local gravitational effects, but rather a global effect on the universe as a whole.

Some scientists have stated that dark energy is the most profound mystery in all of science. Dark energy must exist because something is pushing or pulling our universe apart and overcoming the gravitational forces.

Because of the names (they are both called "dark" because we cannot see them) it's easy to confuse dark matter and dark energy. Though they may in some way be related, their effects are quite different. While dark matter pulls other matter toward itself (like gravity), dark energy pushes the whole universe outward. Also, while dark matter exerts its influence on individual galaxies as well as on the universe at large, dark energy shows itself acting only on the largest cosmic scale over the whole universe.

"The great mystery is not why there is dark energy. The great mystery is why there is so little of it," said Leonard Susskind of Stanford University, at a 2007 meeting of the American Association for the Advancement of Science. "The fact that we are just on the knife edge of existence, [that] if dark energy were very much bigger, we wouldn't be here, that's the mystery." Even a slightly larger value of dark energy would have caused spacetime to expand so fast that galaxies, stars and planets would not have formed,

Is it wise for a physicist to admit that only 5% of our universe is observable, and at the same time say that there is no God? Is it good science to make definitive declarations about the origins of our universe and of life when the vast majority of the energy in our universe is still an unknown entity? Can random chance really account for the existence of our world or for the existence of a single planet?

3.

The world's main building blocks - quarks and leptons

Elementary, my dear Watson.

Sherlock Holmes

Many years ago, in the dark ages, I took a course named Nuclear Physics. I say dark ages, because since then, science has made many amazing discoveries about our world. One thing that has fascinated physicists throughout the atomic age is the question of what holds the nucleus together. Two protons sitting in the same nucleus should blow apart like two north poles of a magnet.

Back then I learned that the atom was the basic building block for all things that exist in our world. The atom always contains at least these three ingredients – a positively charged proton at the center, a negatively charged electron spinning merrily in orbit around the central nucleus, and a neutron that carries no charge, also located in the nucleus.

Protons and electrons are like opposite poles of a magnet. Therefore, they strongly attract each other. Electrons, since they are very much lighter than protons are strongly pulled toward the nucleus. However, since they are spinning at a very high velocity (about 1% of the speed of light), they always miss the nucleus as they try to fall into it.

If there were no proton in the nucleus, the electron would fly off on a straight line, but since it is pulled inward by the attraction to the proton, it falls around the center. This force is similar to gravity that keeps our space satellites in orbit, and that also keeps the moon in place.

We get different chemical elements by adding more protons and electrons to the atom. Hydrogen has only one of each. Helium has two and oxygen has cight. If you want to know how many electrons or protons an element has, just look at the chemical elements chart from basic science class, and you will see that each element has a number. The higher the number, the more complicated things get.

Since I studied nuclear physics, physicists have learned that what I was taught was only a part of the story. To date, over 200 subatomic particles have been discovered, and most of them are made from only 12 basic particles: the 6 elementary quarks, and the 6 elementary leptons.

For want of a more creative name, scientists call the mysterious force that holds all atoms together, the strong nuclear force. This is the binding force that provides the attraction between protons and protons, protons and neutrons, and neutrons and neutrons, and keeps the nucleus of atoms from blowing apart. A nuclear fission bomb is a device that finds a way to break apart the nucleus of atoms and unleash this powerful force.

This strong force binds quarks together in clusters to make the more familiar subatomic particles, such as protons and neutrons. Very unlike a quark, which responds to the strong nuclear force, a lepton is an elementary, half-integer spin particle that does not know the strong nuclear force exists.

There are two main classes of leptons: charged leptons (also known as the electron-like leptons), and neutral leptons (better known as neutrinos). Leptons are not subject to the strong force interaction, but they are subject to the other three fundamental interactions: gravity, electromagnetism, and the weak force. This is why electrons interact with protons and keep the atom electrically neutral.

The strong force originates in a property known as colour. This property, which has no connection with color in the visual sense of the word, is somewhat similar to the forces of an electric charge. Just as electric charge is the source of electromagnetism, or the electromagnetic force, so colour is what responds to the strong force. Particles without colour, such as electrons and other leptons, are not affected by the strong force; particles with colour, principally the quarks, do "feel" the strong force.

Fundamental quark particles come in six different varieties. In order of decreasing mass, they are referred to as: up, down, strange, charm, bottom and top. The up and down quarks are stable and make up protons and neutrons.

Protons are made of two up quarks and one down quark, and neutrons are made of two down quarks and one up quark. Leptons also have six different types –the electron, the electron neutrino, the muon, the muon neutrino, the tau, and the tau neutrino. Our universe is a very complicated but also intricately balanced system.

As scientists dig deeper into the atom, they are finding that protons and neutrons not only have quarks, but also anti-quarks, Higgs-Boson particles and other exotic entities bouncing around inside them.

The study of the atom keeps getting more and more complicated. When it seems like they have finally found the Holy Grail of matter and there is nothing smaller left to be found, they find something new. They uncover layer after layer of complexity. Physicists still don't know why some particles have weight and some do not, or where the weight even comes from.

When I was about six years old, I tried to imitate my dad who sometimes fixed clocks and watches. I found a nice gold pocket watch in one of his drawers and proceeded to take it apart to understand how it worked. Needless to say, I was a little over my head; the farther I went into the watch the more complicated it became and I ended up with many more parts then I knew what to do with.

When I realized that I could not put things back the way they were, I decided to get rid of the incriminating evidence and I buried the watch and its many parts outside the house in a pile of ashes from the furnace. After some time that the watch could not be found, my Dad implored us kids to tell him anything we knew about it. In tears I took him to the ash pile and revealed my folly. It turned out that the watch was borrowed from a friend who was a local jeweler.

What is inconceivable is that some scientists believe that this incredibly complex world came into existence all by itself. As a six-year old, I knew that the watch I was working on didn't just appear out of thin air: it had a maker. Nothing so intricate could ever form by itself, and yet the universe is a million times more complicated than a gold pocket watch.

How did this very intricately arranged universe take shape? Where does the Strong Nuclear Force come from? Where does any force in our universe come from?

Did quarks evolve? If the building blocks of atomic matter were the least bit different, would any world exist? If the proton had a quark of a different colour, would you be here?

4.

A perfect universe

Two things are infinite: the universe and human stupidity; and I'm not sure about the universe.

Albert Einstein

An overview of indispensable conditions for complex life:

The existence of life on Earth is a mystery to science. Since man began to reason, he has wondered where the world comes from and how life came to exist. The common, historical, prevailing explanation for our existence is that some greater power made us.

One of the reasons man believes this is because as he looks at the conditions for life as he knows it, he sees that on Earth everything is "just right" for supporting life. If we were to take away even one of a hundred ideal conditions, our life could not exist. Here are just a few of the ideal conditions that exist in our world:

- If electrons stopped spinning in their orbits around atomic nucleases, mass would cease to exist. Gravity would disappear, electricity would not flow, no chemical re-actions of any kind would take place, and the universe would cease to exist.
- Earth's life-sustaining conditions are possible because of its position in the solar system. It is not too hot and not too cold – if the earth were much closer to or farther from the sun, the planet could not support complex life.
- Even at our ideal distance from the sun, without the proper amounts of water and carbon controlling our temperature (greenhouse effect), the earth would be -73 degrees Celsius, much too cold to support life. Our stable temperature range (-50 to about +50 degrees Celsius) allows life to thrive.
- If Earth did not rotate on its axis, or rotate around the sun in a fairly circular orbit, the temperature extremes would kill all complex life.
- If the composition of Earth's atmosphere varied by more than a few percentage points, most life as we know it would come to an end.

- If the percentage of water covering the earth were to change by only a few percent, life would also change drastically, and most creatures would perish.
- If we didn't have a moon like ours of the right size and distance from the planet, or if the moon's orbit were different, our ecosystems would no longer support life. There would be either no lunar tides, or tides that would be "tsunamis" killing all life along the shores.
- If the earth did not have a heavy rocky core containing high amounts of iron, surrounded by molten rock that flows around the core, there would be no magnetic field. If we had no magnetic field, we would have lost our atmosphere long, long ago and have been bombarded by killer cosmic rays. Magnetic navigation would not have been possible. Birds and whales could not migrate, and bees would never find their way home to make honey.
- If there were no Van Allen belts and other shields high above the earth, all life would have been micro-waved to death eons ago by damaging solar radiation.
- If there were no ozone layer, no one would dare lay exposed on a beach for more than a few minutes, and most life could not exist on the surface of the planet.
- If our planet's rotation were to significantly speed up or slow down, life could no longer be sustained. Temperatures would be too hot or too cold. Wind velocity would damage most vegetation and moisture would not be carried from place to place.
- If Earth's tilt on its axis (23.5 degrees) were changed by 5 degrees, seasons would be so severe that they would kill most plants and animals, and humans would have to migrate every year to stay in habitable temperatures.
- If our solar system were located farther in or farther out in our galaxy, conditions on Earth would not permit life. It would have been sucked into one of the spiral arms or into the dark hole at the center.
- If there were no evaporation and condensation, there would never be any rain, nor clouds, nor fog. Air conditioners would not work, and sweat would keep building up on the skin. Teakettles would not whistle and water would never boil.

- If water did not expand when it freezes, all life in non-tropic rivers, lakes and parts of the ocean would have perished long ago.
- If there were no salt in the sea, it would be a hot-bed for disease. If the salt content were too high, life could not exist (a good example is the Dead Sea - There are no plants, fish, or any other life there. Its salt concentration is a staggering 33.7%, 8.6 times saltier than ocean water).
- If there were no buoyancy, surface water travel would be impossible, and most life forms would have been constrained to the place of their origin. The Hindenburg blimp would never have made it to New Jersey.
- If air pressure did not get reduced at higher speeds, airplanes and birds could not fly and carburetors would never draw in any fuel to the cylinders.
- If there were no capillary action, water would never make it to the places it needs to be to support life, and blood would not circulate in anyone's body.
- If there were no surface tension, neither water drops, nor teardrops would ever form. A liquid spill would run off the table before you could get something to sop it up (using capillary action, of course).
- If there were no chlorophyll in plants, they would not grow, and there would be no food for most living creatures.
- If there were no winds, water would not get redistributed on the earth's surface, Columbus would never have "discovered" America and the Spanish Armada would have never left port.
- If there were no electromagnetic radiation in our universe, life of any kind would be impossible because there would be no heat or light and therefore no energy.
- If carbon were not abundant on our planet, none of the life forms that we know of could have existed.

The above list contains factors required for complex life, but life is not guaranteed to arise even if all these factors are met. This is just a partial list of some of the most obvious ideal conditions for life. Taken separately, each ideal condition is highly improbable by chance, but when taken together, the probability is so small as to be absolutely impossible. The alternative explanation to all these ideal conditions occurring by chance is "intentional design by a supernatural being," a wholly realistic explanation.

Ours is what we might call a "perfect world." Everything is "just right" to support life. We take all of these things for granted, assuming that this is normal, but not realizing what should be obvious – all of these things didn't just happen.

The coincidence of all of these Goldilocks conditions occurring by chance is impossible to the nth degree. Nothing as unlikely as this has ever happened by chance. The best comparison might be if you were to win the jackpot in your state's lottery every day for fifty years. That's how unlikely it is that all of these ideal conditions for life on Earth could have happened by chance.

Did we win the "cosmic lottery"? In the story of Goldilocks and the Three Bears, Goldilocks tried the porridge, the chairs and the beds, and after testing each one, decided that the things belonging to baby bear were "just right."

In the next chapters, we will see that in our world, "everything is just right" as Goldilocks said about Little Bear's things. In checking out the necessary conditions for life on our planet, we will look at many more than just three simple "just rights" and see that if they were not "perfect," life could not exist.

5.

Gravity and other forces

"On the contrary, Watson, you can see everything. You fail, however, to reason from what you see. You are too timid in drawing your inferences."

Sherlock Holmes

Gravity is the mutual attraction between any two masses. This sounds very simple, except not all atomic particles and sub-particles have mass. Physicists are still trying to find out why some have mass and some don't, or why the mass is different.

For centuries, it was believed that there was no force stronger than Gravity. Now it is believed that Gravity is the weakest of the four forces acting on our universe. The four forces are: 1) Gravity (that causes bodies to attract each other), 2) the Strong Nuclear Force (that holds the nucleus of the atom together), 3) the Weak Nuclear Force (that causes radioactive decay), and 4) the Electromagnetic Force (that, among other things, keeps the electrons orbiting the protons in the atom).

The mightiest of the four indispensable forces in our universe is the strong nuclear force, but it only acts over a very short range, about the size of the nucleus of an atom. The weak nuclear force is about one-millionth as strong and has an even shorter range – about the diameter of a proton. The electromagnetic force is about one-hundredth as strong as the strong nuclear force, but its influence is infinite because the photons that carry this force travel at the speed of light and have no distance limitation.

Lastly, gravity is the weakest of the four, being only about 6x10-29 times the strength of the strong nuclear force, but it also has an infinite range of influence. Nobody is quite sure yet where the forces coming from dark matter and dark energy fit in.

Gravity is the force that gives objects weight. The weight of an object is determined by how many protons, neutrons and electrons an object contains. Earth's gravitational field holds the atmosphere in place preventing oxygen, nitrogen, carbon dioxide and other gases from escaping into space. It also exerts such a potent force on the core of our planet that it turns seething, molten iron into solid rock. Without gravity there would be no planets or stars. It also keeps our oceans and Earth's surface plates from floating off into space.

Thanks to Newton's nearly universal law, we can figure out the gravitational force between any two objects. We take it for granted now, but more than 300 years ago Sir Isaac Newton proposed a revolutionary idea: that any two objects, no matter their mass, exert gravitational force toward one another. This law is represented by an equation that many high school students encounter in physics class. It goes as follows:

$$F = G \times [(m1 \text{ x } m2)/r^2]$$

F is the gravitational force between the two objects, measured in newtons: m1 and m2 are the masses of the two objects, while r is the distance between them. G is the gravitational constant, a number calculated to be $6.674 \times 10\ {-11}\ N{\cdot}m^2/kg2$.

The benefit of this formula for the universal law of gravitation is that it allows us to calculate the gravitational pull between any two objects. This ability is especially useful when scientists are planning to put a satellite in orbit, charting the course of the moon, or trying to determine if an asteroid is going to strike the earth.

As mentioned earlier, mass is only the felt effect of energy interacting with other energy, and the force of gravity depends on that energy. In fact, the whole world is made up of nothing but energy composed of positive and negative electrical charges and of particles that inherently attract each other.

When we add up the total energy force of a clump of electrons, protons and neutrons we can predict what attraction (gravity) it will have to another clump of electrons, protons and neutrons. The weight of any object is determined by the force of attraction between that object and any other object. The shorter the distance between the objects, the stronger the force. Also, the greater the number of particles of energy in each object, the greater the force.

Every body in the universe is attracted to every other body in the universe, but since distances are usually very large, that force is seldom felt. Our human bodies are attracted to the sun, and to the moon, and to the other planets in our solar system, but we don't physically feel anything. We are also attracted to the most distant stars. Our tides occur because our oceans are attracted to the moon and the Sun.

Since we are in such close proximity to the earth and it is so massive compared to anything else near us, the only gravitational force we can feel is from Earth itself.

It has been said that when a mother draws her child close to her breast, she is exerting a gravitational pull a million times stronger than the pull of the moon. Maybe that's why everyone needs a hug, and group hugs are even better.

Could our universe exist if there were no attraction between energy particles bundled together in the form of matter? What would happen to you if there were no gravity? What would happen to our vital atmosphere if Earth's gravity did not keep it here?

Who regulated the force of Gravity to be just slightly less than the force of dark energy in the universe? Who invented this perfectly balanced tug-of-war that allows objects in our universe to exist?

6.

Bang, bang, you're alive

"Nothing comes from nothing: nothing ever could."
Maria, in the Sound of Music

The big bang theory -

Based on research performed by Edwin Hubble, Georges Lemaitre and Albert Einstein, among others, the big bang theory postulates that the universe began almost fourteen billion years ago with a massive expansion event. At the time, the universe was confined to a single point, containing all of the universe's content. That original outward impulse continues today, as the universe keeps expanding slowly, but at a slightly increasing rate.

The theory of the big bang gained widespread support in the scientific community after Arno Penzias and Robert Wilson discovered cosmic microwave background radiation in 1965. Using radio telescopes, the two astronomers detected cosmic noise, or static, that didn't dissipate over time. Collaborating with Princeton researcher Robert Dicke, the pair confirmed Dicke's hypothesis that the original big bang left behind low-level radiation detectable throughout the universe.

If the big bang theory is true, we still need an explanation of where the energy came from. The problem of the origin of the mass that exploded and the space that exploded with it is the giant elephant in the scientific room that most researchers try to ignore. So far, science can only hazard a guess about where the substance of the universe came from. It can't explain the origin of one single atom or what provides the force that holds its nucleus together.

There are still many unanswered questions about how the big bang could work. For instance, when you have a normal explosion, it creates a vacuum at its center where nothing exists for a period of time. However, when we look around us at our universe, we see no zone where there is nothing. Some have hypothesized that the universe is like the skin of a balloon that is expanding as it is blown up. But, in a balloon, you have an empty space at its center.

As we peer outward from our planet into space, we see that everything, on every side around us, is moving away at the same velocity. That fact would seem to put us at the center of the universe, but could not explain how we have stayed here after the explosion. There are so many things that we still do not understand about how the universe works.

So far, no one has provided a reasonable explanation for the origin of mass and space in our universe unless they have admitted the existence of a greater being that is outside of this physical realm. Even if you should agree with Richard Dawkins (world renowned, atheistic professor at the University of Oxford) and suggest that life on our planet was seeded here by aliens, you could not be so naive as to believe that aliens from outer space also created the universe.

One law of nature that I have never seen stated clearly is that: "Nothing in our universe exists without a maker." If you look around you at a building, or a car, or a paved road, or a suit of clothes, you know without a shadow of a doubt that they were made by someone. Why should you think differently when you look at a tree or a sparrow or a galaxy that are many times more complicated?

The Big Bang theory tells us that there was not only a beginning of the universe in time, but there was also a beginning of time itself. However, some of those who believe in "a beginning" do not want to admit that, if there was a beginning then there must of necessity be a Beginner; someone responsible for creating the elements we see today.

7.

An expanding universe

They say the universe is expanding. That should help with the traffic. Steven Wright

Let's look more closely at Edwin Hubble, after whom the Hubble telescope was named. While the 1920s were roaring past and then as the Great Depression shuffled by, Hubble was performing groundbreaking astronomical research. Hubble not only proved that there were other galaxies besides the Milky Way, he also discovered that these galaxies were moving away from our own, a motion he called recession.

In order to quantify the velocity of this galactic movement, Hubble proposed Hubble's Law of Cosmic Expansion, known as Hubble's law, an equation that states: velocity = $H_0 \times$ distance. Velocity represents the galaxy's recessional velocity; H_0 is the Hubble constant or parameter that indicates the rate at which the universe is expanding; and distance is the galaxy's distance from the one with which it's being compared.

The way to calculate Hubble's constant has varied over time, but the current accepted value is 70 kilometers/second per mega parsec (a unit of distance in intergalactic space). Hubble's law provides a precise method for measuring a galaxy's velocity in relation to our own. Most importantly, the law established that the universe is made up of many galaxies whose movements trace back to the big bang.

Two factors control how fast the universe has been expanding: 1) its mass density and 2) its space energy density. These two factors must be precisely fine-tuned in order for the universe to expand at the correct rate. Anything different, and life would not be possible anywhere in the universe.

If the universe had expanded even slightly faster than it did, gravity wouldn't have been strong enough to permit galaxies to form. Without galaxies, and specifically the just-right type of galaxy, life would not be possible.

On the other hand, if the universe had expanded even slightly slower than it did, the universe would have collapsed under the effects of its own gravity before any stars could have formed.

Without stars, and specifically a just-right star with a just-right solar system, life would not be possible.

Some scientists say that if the mass density of the universe varied by more than even one part in 10^{60}, no heavenly bodies would have formed, and no life would be possible. John Jefferson Davis has pointed out that having an accuracy of one part out of 10^{60} is like "firing a bullet toward the other side of the observable universe, twenty billion light years away, and hitting a one-inch target!"

The incomprehensibility of this extreme degree of fine-tuning is more fully appreciated when one considers that there are only about 10^{25} grains of sand on all the beaches on Planet Earth. Fine-tuning to the order of 10^{60} would be many times more exact than saying that not a single grain of sand could be added to or removed from one beach on Planet Earth without causing the earth to be unable to support life.

Because the mass density and space energy density of the universe are in perfect proportion, the universe has expanded at the precise rate needed to permit the right objects to form and therefore allow life to exist. By the way, a single grain of sand has more atoms in it than there are stars in the Milky Way.

Who got the velocity of expansion just right? Why do the forces of gravity and the forces expanding the universe differ just enough to allow the formation of galaxies, stars and planets?

Section II –

How Earth won the lottery

On the Science Channel, spokesmen often make statements like: “fortunately for us,” or “we have had the remarkably good luck,” or “it’s our good fortune that,” when referring to certain favorable conditions in our universe.

As scientists consider some of the known conditions and laws that exist in our universe, galaxy, solar system and planet, they must all admit that we are very fortunate to have such a habitable environment that favors and promotes life. If there were such a thing as The Cosmic Powerball Lottery, Earth would have taken the jackpot.

Naturalistic evolutionists say that regardless of what world might have appeared, life would have adapted to it. They say that if there were no atoms, the world would be made of croutons or some other basic elements. If there were no oxygen, critters would breathe something else. If there were no plants, complex life would have found something else to eat, and so on. But these assertions are based on highly speculative thinking and are not science. Science is the study of the real world and not an imaginary one.

The truth is that the world we live in is the only world we know to exist, and this world has all of the ideal conditions needed to support life. No matter where we have looked in the heavens, we have not found a place that has all of the favorable and necessary conditions to support life like Earth has.

8.

Let there be light

The light shines in the darkness, and the darkness can never extinguish it. John, The Apostle

We can easily forgive a child who is afraid of the dark; the real tragedy of life is when men are afraid of the light.
Plato

Light is essential for most if not all of life to exist. Light from the sun is what gives life to all of Earth's plants that provide food for various levels of the food chain in the circle of life. The major reason that light is indispensable is because light causes heat and heat is indispensable for life.

A light wave is an electromagnetic wave of a certain frequency range that travels through the vacuum of outer space. Light waves are produced by vibrating electrical charges. An electromagnetic wave is a transverse wave that has both an electric and a magnetic component that are perpendicular to each other.

All planets in our solar system receive light from the Sun, but no planet uses it as effectively as Earth. Trees and plants on Earth produce oxygen through a process called photosynthesis. Plants need the light to grow.

Look at plants in windows and notice how they always lean towards the Sun. Try growing one plant in a dark room and another in a light room; notice which one grows slower, if at all. The one that has grown fastest is the one that also produces more oxygen.

One necessary factor for plants to survive and produce oxygen on Earth is the length of time the planet takes to spin once on its axis. The revolution of 24 hours means that each side of the planet receives sunlight regularly. If we look at a planet like Venus, that takes 243 days to spin on its axis, it means that for a large period of time certain parts of the planet are in complete darkness.

Even if that planet could support life of some kind, it would struggle to do so. Almost all plants like those on Earth would die before their side of the planet turned to face the sun again.

The sunlight has many other beneficial properties. Each human cell and every form of life on Earth synchronize their activity with the sun. Hormones, neurotransmitters, inflammatory and growth factors, self-synchronized genes and many more yet-to-be discovered mechanisms regulate the functioning of living organisms.

What all of them have in common so far is that their life cycle is regulated on a daily, monthly or seasonal basis. In other words, they are dependent on the availability of sunlight at regular intervals.

At the beginning of each period of sunlight, light-sensitive central nervous system compounds launch hundreds of reactions in human beings. The number of monoamines, the nervous system's most important neurotransmitters, increases.

At the same time, hormonal regulation in different areas of our bodies is synchronized through chemical communication according to this rhythm. Successful synchronization of these activities is a prerequisite for our performance and well-being.

Decreased amounts of light or complete darkness activate "night hormones" like melatonin, whose level sharply increases at night and reaches its peak in the early hours of the morning. Exposure to artificial light in winter or during the evenings can negatively affect well-being by interfering with the essential rhythm of metabolic events. Nausea and fatigue easily increase in artificially lit environments due to the lack of natural and bright light signals. That is why people who live year-round in polar regions often suffer from insomnia and other physical and mental problems, especially during the long weeks of winter darkness.

Shifting sleep schedules affect our body's circadian rhythm. Disrupting it can throw everything out of balance -- including our cardiovascular system, metabolism, digestion, immune system, and hormonal balance. The list of negative consequences is quite long. There is an increased risk of diabetes, obesity, breast cancer, negative metabolic changes, heart attacks, physical injury, and depression.

It is known that shift workers' disrupted body rhythm, that manifests itself in abnormalities of melatonin secretion among other things, predisposes them to a larger number of cancers. Frequent changes in day and night rhythms are known to cause wide-ranging problems in metabolism overall: it disturbs blood glucose control, worsens fat levels, and increases blood pressure, obesity and mood swings.

The need for light for us is usually self-evident, though we often underestimate its central role in our own well-being and for nearly all of life on our planet.

Having a regular daily exposure to sunlight is important for life to thrive on planet Earth. Have you thought about the importance of light for life? Who lit the match to turn on the light of our Sun? Would we be here if our Sun had never gotten "turned on?"

Would life have developed anyway, in whatever conditions were available as some naturalistic evolutionists seem to claim? Could life really be based on darkness instead of light?

9.

Heat from the Sun

Heat cannot be separated from fire, or beauty from the eternal. Dante

Some people change their ways when they see the light; others when they feel the heat. Caroline Schoeder

We are told that at absolute zero (-459° F/-273° C), molecules cease to move. Nothing can live if there is no molecular movement. Earth has a suitable climate for sustaining life because it has heat that keeps molecules moving. Most of life's heat comes from the sun, but some of it is also due to the heat left over from the earth's formation.

The temperature on Earth rarely goes rapidly from very hot to very cold or vice versa. In contrast, the surface on the planet Mercury can be anything from 200°C below freezing to 375°C above. At 375°C, water would only exist as a gas, and the planet would be completely dry.

Venus has a surface temperature of 480°C that would be much too hot for anybody to live on. Mars, although it can reach a maximum of 25°C, can be as cold as -140°C, a temperature that would freeze alcohol, gasoline, blood, water, and many other liquids. The other planets in our solar system are colder still.

If we had no Sun, there would be no life on Earth. Without the sun's gravity, the planet Earth wouldn't be here either; it would be floating aimlessly through the coldness of interstellar space. Because of Earth's ideal distance from the sun, it receives the perfect amount of heat and light to allow life to flourish. Imagine what would happen if the sun suddenly vanished or its nuclear fire went out. How would you keep warm? How would you see? How would you get food and drink? How would plants and trees grow? How would they photosynthesize?

Even though other stars give us tiny pin points of light, that light is insignificant compared to the light that comes from the sun. Only the sun is close enough to bring light and life to our world. Our planet is one of the very few places in the universe that is close enough to one star that its whole surface is bathed in light, but at the same time far enough away that life is not destroyed by the heat.

The infrared rays of the sun give us the warmth we need to live and the visible spectrum of sunlight enables us to see, and also provides for photosynthesis in plants. Light is just a tiny part of the electromagnetic spectrum. You could not selectively eliminate just the frequencies of visible light from the universe without eliminating all the other forms of electromagnetic radiation at the same time.

There are a few plants and minute organisms that seem to exist without sunlight, but they cannot exist without the heat that the sunlight continually renews on our planet.

Alligators and crocodiles get 80% of their energy directly from the sun. Crocodilians are ectotherms, producing relatively little heat internally and relying on external sources to raise their body temperatures. The sun's heat is the main means of warming for any crocodilian. These animals cannot exist in cold climates and could not exist if our nights were twice as long.

No heat equals no life, and our heat source turned out to be nearly perfect with built-in atmospheric thermostats that are incredibly effective at maintaining our planet's temperature. Who was the engineer that installed this complex system of temperature controls? Who placed us in a perfect orbit at just the right distance from our abundant source of heat and light?

10.

The Sun's type and location in the galaxy

"It's a mystery. That's the first thing that interests me about the idea of God. If there is one, it's mysterious and powerful and awesome to even consider the concept, and you have to take it seriously." Author Stephen King

For life, we must have the right star

The sun is an ideal type of star for our needs. It is steady burning, long-lived, and neither too large nor too hot. The vast majority of the stars in our galaxy are much smaller than our Sun and provide neither the right kind of light nor the right amount of heat to sustain life on an earthlike planet.

In addition, most stars are gravitationally bound to one or more other stars and revolve around one another. Our Sun, by contrast, is independent. It is unlikely that our solar system would remain stable if we had to contend with the gravitational force of two or more stars.

A closer look at our "jackpot" Sun - a spectral type G2, dwarf, main-sequence star -

- Total number of stars estimated in the Milky Way – 100-200 billion
- Over 80% of these are low-mass red dwarfs (these most likely lack a habitable zone due to a lack of heat and light at a safe distance)
- If our star were smaller, the size of the habitable zone would shrink and any planets in that zone would be too close and would become tidally-locked into a synchronous orbit (1 rotation = 1 revolution), as our moon is with Earth
- 1-2% of the stars are massive short-lived blue giants that could not support life
- Only about 4% of the stars are early G-type, main-sequence stars like our Sun
- A high percentage of those are not in the Galactic Habitable Zone where life could exist for an extended period

- 50% of the G-type are in binary systems – two stars at the center which condition creates erratic planetary orbits

This means that less than 1% of all of the stars in the Milky Way give any hope of supporting life.

The right location in the right kind of galaxy

The Milky Way is neither round nor homogenous. Instead it is shaped like a spinning octopus with various trailing spirals. Our solar system is located between two of these arms in a region that has relatively few stars. Nearly all the stars that we can see at night are so far from us that they remain mere points of light when viewed through the largest telescopes.

Scientists believe that the sun and our Solar System have been located in a stable orbit within our galaxy for the last 4.5 billion years. We are far from the deadly center of our galaxy and firmly located between two of its dangerous spiral arms. The stability of our position is also possible because the sun is one of the rare stars that lies within the "galactic co-rotation radius."

The vast majority of the stars in our galaxy orbit the center of the galaxy at a rate that differs from the rate of the trailing spiral arms. Thus, most stars located between spiral arms do not remain there for long, but are eventually swept inside one of the arms that are jam-packed with many stars and other matter.

Only at a specific distance from the galaxy's center, the "co-rotation radius," can a star remain in its place between two spiral arms, orbiting at precisely the same rate as the galaxy arms rotate around the core.

Why is it important that we are not within one of the spiral arms?

1) First, being outside the spiral arms puts us in a location that is safer from collisions, radiation and gravitational influence than anywhere else we might choose in the universe.

2) Second, we are removed from the more densely occupied areas, where stellar gravitational forces can lead to disruption of planetary orbits.

3) Third, we are farther from the deadly effects of supernovae explosions.

4) Fourth, the 4+ billion-year stability permitting life on Earth would not have been possible at most other locations in our galaxy.

5) And fifth, our location gives us a view of the universe that is unobstructed by the stars, debris and gases found in the spiral arms. If we were within the spiral arms, our view would be significantly impaired.

If our solar system were close to the center of the Milky Way, it would suffer the harmful effects of that dense concentration of stars. Earth's orbit, for example, would likely be disturbed, and that would dramatically affect the possibility of life.

As it is, our solar system appears to have just the right position in the galaxy to avoid this gravitational tug-of-war and other dangers, such as overheating when passing through gas clouds or being exposed to exploding giant stars and other sources of deadly radiation.

The dense centers of other galaxies such as NGC 7331 (often referred to as a "twin" of the Milky Way) have high levels of radiation that are dangerous to complex life. The *Rare Earth* hypothesis suggests that much of the known universe, including large parts of our galaxy, cannot support complex life. Those parts of a galaxy where complex life is possible make up the "galactic habitable zone." This zone is primarily a function of distance from the galactic center. As that distance increases:

1. Star metallicity declines. Metals are necessary to the formation of terrestrial planets.
2. The X-ray and gamma ray radiation from the black hole at the galactic center, and from nearby neutron stars, becomes less intense. Radiation of this nature is considered dangerous to complex life, hence the *Rare Earth* hypothesis predicts that the early universe, and galactic regions where stellar density is high and supernovae are common, would be unfit for the development of complex life.
3. Gravitational disturbances from planets and planetesimals (objects formed from dust and other debris) from nearby stars becomes less likely as the clustering of stars decreases. Hence the further a planet lies from the galactic center or from a spiral arm, the less likely it is to be struck by a large bolide that

is an exceptionally bright and powerful fireball. A sufficiently large impact might extinguish all complex life on a planet.

The first item rules out the outer reaches of a galaxy, the second and third rule out galactic inner regions, globular clusters, and the spiral arms of spiral galaxies. As one moves from the center of a galaxy to its furthest extremity, the ability to support life rises then falls. Hence the galactic habitable zone is probably ring-shaped and sandwiched between its uninhabitable center and outer reaches.

Rare Earth proponents conclude that the solar system of a life-enabling star must also have a galactic orbit that is nearly circular in its journey around the center of its galaxy. The required synchronization of the orbital velocity of a central star with the wave velocity of the spiral arms can occur only within a fairly narrow range of distances from the galactic center. Lineweaver and others calculate that the galactic habitable zone is a ring 7 to 9 kilo parsecs in diameter, and that it includes no more than 10% of the stars in the Milky Way.

Based on conservative estimates of the total number of stars in the galaxy, this could represent something like 20 to 40 billion stars. Gonzalez, and others would halve these numbers; he estimates that at most 5% of stars in the Milky Way fall in the galactic habitable zone, especially when you account for the complicated issue of the spiral arms.

Being in the right place for the right time

While a planetary system may enjoy a location favorable to complex life, evolutionary scientists believe that it must also maintain that location for a span of time sufficiently long for complex life to develop. A central star with an orbit that steers clear of galactic regions where radiation levels are high, such as the galactic center and the spiral arms, would appear necessary.

If the central star's galactic orbit is eccentric (elliptic or hyperbolic), it will pass through some spiral arms, but if the orbit is a near perfect circle and the orbital velocity equals the "rotational" velocity of the spiral arms, the star will drift into a spiral arm region only very gradually - if ever.

The orbit of the sun around the center of the Milky Way is indeed almost perfectly circular, with a period of 226 Ma (1 Ma

= 1 million years), one closely matching the rotational period of the galaxy.

The Andromeda and the Milky Way galaxies have similar masses, but whereas Andromeda is also a typical spiral galaxy, the Milky Way, by comparison is unusually quiet and dim. It appears to have suffered fewer collisions with other galaxies over the last 10 billion years, and its peaceful history may have made it more hospitable to complex life than galaxies that have suffered more collisions, and consequently more supernovae and other disturbances. The level of activity of the black hole at the center of a galaxy is also important. The Milky Way's black hole appears to be just right.

Our planet, our star, our solar system and our galaxy couldn't be made any better. It's difficult to imagine how we could improve on their design. If someone were going to make a perfect world, astronomically speaking, it would look exactly like ours.

11.

A terrestrial planet of the right size

"The mystery of the beginning of all things is insoluble by us; and I for one must be content to remain an agnostic."
Charles Darwin

One size does not fit all where life is concerned. A planet that is too small cannot hold much of an atmosphere. Therefore, the surface temperature varies much more and the average temperature drops. Substantial liquid oceans become impossible. A small planet will also tend to have a rough surface, with large mountains and deep canyons. The core cools faster, and plate tectonics (the floating of numerous solid surface plates on top of a plastic or liquid base) will either not last as long as they would on a larger planet or may not occur at all.

A planet's atmosphere depends very much on what composition of gases was present when the planet was formed. Did it have oxygen, carbon dioxide, argon, helium and hydrogen and in what quantities?

All planets with atmospheres lose some of their contents, especially the lighter gases, to outer space. The degree of this loss depends on planet mass, its atmospheric composition, and its distance from its star.

Hydrogen is the lightest of all elements and has a tendency to fly off into space because gravity exerts little force upon it. It, and other gases in lesser quantities are lost to space when the gravity of the planet is low enough for the gas's thermal activity to cause it to accelerate to a breakaway velocity.

Very large planets do not allow this to happen due to their strong gravitational pull. If Earth were twice the size it is, our atmosphere would be very different and probably would not support life. If Earth were as large as Saturn for instance, its atmosphere would be comprised mostly of hydrogen.

For life to exist, it requires rocky planets like Earth with a solid surface. Since gas giant planets lack such a surface, or have it so deeply immersed in its atmosphere that the pressure is crushing, complex life cannot exist there. The two very large planets in our solar system (Saturn and Jupiter) are comprised almost completely of gases and do not have solid rocky surfaces. If they do have solid cores, they are very, very small in relation to their total size, and may consist only of gases like hydrogen that are compressed into a solid state by the crushing force of gravity.

NASA's space probe JUNO (launched May 2011 and recently arrived) will try to determine if indeed Jupiter has a solid core and what its size and composition might be. The atomic composition of Jupiter is 90% hydrogen and the rest is mostly helium, with a trace of other elements. A giant planet like Jupiter is like a black hole for gases like hydrogen and helium. They are trapped there and cannot escape the strong forces of gravity.

The closer a planet is to its star(s), the more it heats up and the more it suffers from solar storms that propel ions outward expelling part of the planet's atmosphere into space. There is a continual bombardment that fluctuates according to sunspots and storms on the solar surface. Gravity pulls the gases down toward the center of the planet while heat causes the gases to expand, to accelerate and then to try to escape this gravitational pull.

There is a constant tug-of-war going on between the planet's gravity and the velocity of the elements due to heating. Therefore, to sustain life, a planet must be neither too close nor too far away from its star, or the atmospheric conditions will not support life.

So, in order to have a planet with a habitable atmosphere, we need one that is 1) not too large and not too small, 2) that is not too close or too far from its star, and 3) that began with the right gases in its composition.

Congratulations Earth: You won first prize in the size category.

12.

Earth is "just right"

According to Goldilocks, little bear's chair was just right. According to many scientists, Earth is just right for sustaining life. We have seen that a "habitable" planet must:

- Orbit a star that remains stable in output for billions of years
- Be at a distance from the star that results in it achieving a suitable temperature so its surface water is mainly liquid, not frozen
- Have a nearly circular orbit, so that constant conditions prevail for its entire "year"

These requirements lead to the concept of a "habitable zone" within a solar system, similar to the galactic habitable zone we already discussed for the Milky Way. Not all stars can support a planet with life. A low mass star that is significantly smaller and cooler than our Sun could only sustain life on planets very close to it, while higher-mass, hotter stars make their close-in planets too hot and so life could only be sustained farther out.

Stars more than about twice the mass of the sun would not last long enough for life to form by random evolution on their planets. If a planet is too close to a star, its spinning will synchronize with its orbital motion and it will always point the same side to the star - it will be "tidally locked," like the planet Mercury is nearly locked to the sun with the same side nearly always facing it. When this happens to a planet, one side is always extremely hot and the other extremely cold.

To this short list, we can add many more general requirements for life to exist on a given planet. For example:

- Must not orbit a star that is close to a cosmic explosion like a supernova
- Be far enough from massive planets that they do not continually divert asteroids to hit it or disturb its orbit
- Not be so massive that it retains hydrogen and becomes a "gas giant"

- It is also essential to have a massive planet like Jupiter well outside its orbit, but still close enough to divert potential devastating asteroids away from the earth.
- Have an atmosphere that is neither too thin nor too thick. Massive atmospheres trap more heat near the planet's surface due to the absorption by certain gases of the sun's infrared radiation. If a planet is close to its star, this extra heating can become excessive, leading to the evaporation of too much water which then produces more carbon dioxide in the atmosphere that increases the heating and leads to what is called a "runaway greenhouse effect," like exists on Venus.

  If the planet is at just the right distance and its atmosphere has just the right thickness, the greenhouse effect can just heat it up mildly and make its climate pleasant, like on Earth.
- Large amounts of water must be available on and near the surface of the planet for many reasons. Water is essential for the chemical reactions leading to and sustaining life. Water also probably plays a role in forming the relatively low-density rocks that make continents that can "float" over the liquid core and plastic mantle and thus create plate tectonics.
- Smaller planets cool relatively fast, and their cores solidify. Once that happens, the planet no longer can drive large scale motions of continents by plate tectonics.

These are just some of the conditions that had to be present for life to exist on Earth. There are many others, some of which we will look at in following chapters. Some scientists (astronomer/astrophysicist Hugh Ross, for one) have made lists of over 128 precise conditions necessary for life to exist and to be sustainable.

Many scientists (like Ward and Brownlee in their book "*Rare Earth*"[1]) believe that the complex life that we find on Earth is very probably the only complex life in the whole universe. That is because it is extremely unlikely that all of the favorable

---

[1] Rare Earth, Ward and Brownlee, Copernicus , New York, 2003, 338 pages

conditions found on our planet, our solar system and our galaxy could be duplicated by chance in any other place. The odds against it are much greater than the number of possible planets in the universe.

Who arranged all of these "dominoes" in a perfect row? Would the whole row of dominoes still fall down if one or two were missing?

13.

Earth's rotation

"Round and round she goes, and where she stops nobody knows" Ted Mack on the Original Amateur Hour

"Tomorrow is the most important thing in life. Comes into us at midnight very clean. It's perfect when it arrives and it puts itself in our hands. It hopes we've learned something from yesterday." John Wayne

Earth's rotation on its axis and its nearly circular orbit around the sun help distribute heat evenly around the planet. If the earth did not rotate, the side facing the sun would be much too hot to support life, and the side facing away from the sun would be much too cold. If it rotated too slowly, the days and nights would be too long and the atmosphere and the surface would alternately overheat or chill so much that most life could have never formed or would die.

Since the earth is composed of solid plates floating on a hot, pliant mantle that floats on a molten sea of liquid rock, if it rotated too rapidly, it would cause it to bulge significantly at the Equator. A higher rate of rotation would cause the liquid part to bulge more where it is moving fastest, and flatten even more at the poles. This would lead to severe volcanoes and earthquakes as the tectonic surface plates were forced out of position.

Because of centripetal force (like throwing a stone from a sling – the faster the rotation the greater the outward force), a rate of rotation twice as fast as we have now would make us four times lighter (some of us might welcome the weight loss) and a rate of rotation ten times faster would make us 100 times lighter. However, the higher rates would also "sling" much of our atmosphere into space.

If the earth had a rotational period of only eight hours instead of twenty-four, we would have surface winds of 500 miles per hour. The highest wind speed ever recorded in a hurricane was 211 miles per hour in Hurricane Gustav in 2008. Very little life could exist facing winds at 500 MPH. Even the life in the water would be perturbed by the tremendous waves and agitation of the seas. All flying creatures would be swept away, and all creeping creatures could only survive underground. No plants would survive so there would be no food. Our atmosphere would be one gigantic sandstorm.

Earth's graceful 24-hour rotation rate is one of the traits that make our planet so friendly to life, allowing most parts of Earth to stay a nice, comfortable temperature as they are bathed in sunlight during the day and near darkness at night.

Each planet in the solar system has its own unique rotation rate. Little Mercury, sizzling closest to the sun, takes 59 Earth days to turn around just once (for all practical purposes it is locked to the sun and suffers severe extended temperature extremes).

Venus, the second planet, rotates once every 243 Earth days. What's more, Venus rotates backwards from the direction of its orbit around the sun, as do Uranus and tiny dwarf planet Pluto. Uranus even lies down on the job, rolling around with its axis of rotation pointed nearly toward the sun. None of these planets could support any complex life similar to what we find on Earth.

The earth's orbit is slightly elliptical (oval). When the earth is closest to the sun (perigee) the southern hemisphere is enjoying summer. The sun heats up land masses more rapidly than the oceans, and since the earth has the bulk of its continents in the northern hemisphere, the whole earth heats up about the same in southern summer as in northern summer. The south's lack of land mass is compensated for by its being 3 million miles closer in summer (in northern summer, it is 94.5 million miles from the sun, and in northern winter it is only 91.4 million miles).

Water in the oceans serves to help distribute the sun's heat around the globe, and because the ocean currents carry more heat from the south to the north than vice-versa, the average temperature in the northern hemisphere is about 2.25 degrees F. (1.25 degrees C.) warmer than in the south. That is a very modest difference, thanks to the fact that the earth's orbit is so close to being circular.

Though Earth's orbit is not a perfect circle, it is only *slightly* elliptical. According to Dr. J. E. Chambers, "computer simulations of planetary formation yield Earth-like planets with large eccentricities (e = 0.15)," whereas the earth has an e value of only 0.03. He goes on to say, "Given that climate stability may depend appreciably on "e," it could be no coincidence that we inhabit a planet with an *unusually* circular orbit."[2]

Who decided that the earth should rotate, and at the perfect velocity that it does? What would happen if Earth were to stop rotating? Would you survive if our planet rotated much faster or slower than it does? Could complex life exist if our orbit were very elliptical? What if the northern hemisphere were at the perigee (closest to the sun) in summer?

---

[2] Chambers, J. E. 1998. How Special is Earth's Orbit? American Astronomical Society, DPS meeting #30, #21.07

14.

Flying through the air with the greatest of ease

Sometimes, flying feels too godlike to be attained by man. Sometimes, the world from above seems too beautiful, too wonderful, and too distant for human eyes to see.
Charles A. Lindbergh

Would you believe it if I told you that in the last 24 hours, you traveled more than 20,000 miles? Did you know that most people on the earth live their lives traveling faster than the speed of sound? On Earth, the speed of sound at sea level — assuming an air temperature of 59 degrees Fahrenheit (15 degrees Celsius) — is 761.2 mph (1,225 km/h), but the speed of Earth's surface at the equator is *1,040 mph.*

Every day, the earth revolves once, and since the circumference of the planet at the equator is 24,901 miles (40,075 km), give or take a few, that means that people who live at the equator are traveling more than 1000 miles per hour. People who live farther south or north are traveling at decreasing rates depending on the circumference of the globe where they live. At the North Pole, the polar bears when sleeping are not moving, but are slowly turning.

We don't realize that we are traveling at all because our atmosphere and all our surroundings turn with us, but the truth is that we are spinning through space at an alarming rate of speed. It's about equal to the speed of two jetliners crashing head-on into each other at cruise velocity. Anything that crashes at that speed is immediately turned to useless trash.

To be exact, the earth rotates once in a few minutes under a day (23 hours 56 minutes 04.09053 seconds). This is called the sidereal period (that means the period relative to stars). The sidereal period is not exactly equal to a day because by the time the earth has rotated once, it has also moved a little in its orbit around the sun, so it has to keep rotating for almost another 4 minutes before the sun seems to be back in the same place in the sky that it was in exactly a day before.

We have already seen how important the planet's rotation is to creating nights and days with alternating cold and heat, and dark and light that are all essential to keeping things on Earth alive.

To this high velocity due to the earth's rotation, we need to add the speed at which the earth is spinning around the sun, and the speed at which the sun is spinning around the center of the Milky Way Galaxy, and the speed at which the Milky Way is also spinning through space. We are hurtling along at an unimaginable velocity.

What would happen if the earth stopped revolving and became frozen in position like the planet Mercury? What would things be like if it had never revolved? Why is our rotation rate just right for supporting life?

15.

Why do we have seasons?

I trust in nature for the stable laws of beauty and utility. Spring shall plant and autumn garner to the end of time.
Robert Browning

As the earth spins on its axis, producing night and day, it also moves about the sun in its slightly elliptical (elongated circle) orbit that requires about 365 1/4 days to complete. The earth's spin axis is tilted with respect to its orbital plane. This tilt is what causes the seasons. When the earth's axis inclines towards the sun, it is summer for the hemisphere on that end of the axis. When the earth's axis points away, winter takes place.

Since the tilt of the axis is 23 1/2 degrees, the North Pole never points *directly* at the sun, but on the summer solstice it points as close as it can, and on the winter solstice as far away as it can. Midway between these two times, in spring and autumn, the spin axis of the earth points 90 degrees away from the sun and both poles are perpendicular to it. This means that on that date, day and night have about the same length: 12 hours each, more or less.

Why should this tilt of the earth's axis matter to our weather? To understand this, take a piece of paper and a flashlight. Shine the light from the flashlight straight onto the paper, and you will see an illuminated circle. All the light from the flashlight is in that circle. Now, slowly tilt the paper, so that the circle elongates into an ellipse. All the light is still in that ellipse or oval, but it is spread out over much more paper. The number of square centimeters increases while the total amount of light stays the same, meaning that there is less light per square centimeter. Also, the light will be slightly dimmer at a greater distance from the bulb.

The same is true on the earth. When the sun is overhead, the light is falling straight on you, and that means that more light (and more heat) hit each square centimeter of the ground where you are standing. When the sun is lower in the sky, the light gets more spread out over the surface of the earth, and less heat can be absorbed.

Since the earth's axis is tilted, the sun is higher when you are on the part of the earth where the axis points more towards the sun, and lower on the part of the earth where the axis points away from the sun.

For the Northern Hemisphere, the axis points most toward the sun on June 22nd, and away from the sun on December 22. This corresponds to the Winter and Summer Solstice. For the Southern Hemisphere, this is reversed.

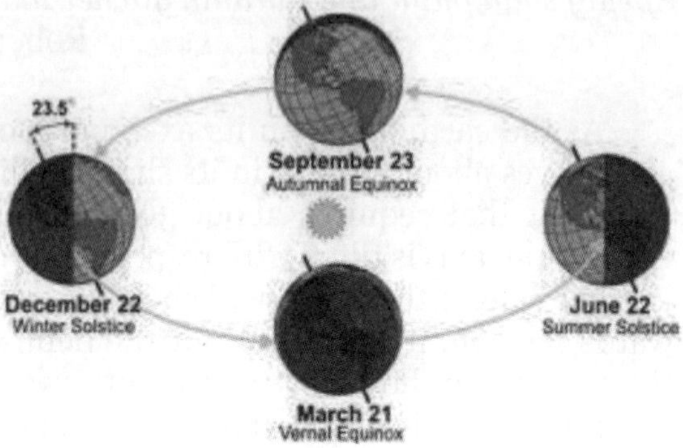

Figure 1-Science.howstuffworks.com

Although the length of the daytime period is always twelve hours at the Equator in all seasons, at all other latitudes the length varies with the season. During the winter, the daytime period is shorter than twelve hours; during the summer, it is longer than 12 hours due to indirect lighting that bends around the atmosphere.

The tilt angles of different planets in our Solar System range from zero for Mercury to 90 degrees for Uranus. Both extremes produce drastic temperature differences across the planet. Mercury, close to the sun and nearly having the same face toward the sun all the time, ranges from extreme heat (800 °F) at the equator to ice at the poles.

Not only is the earth's angle of tilt at the optimum value, it also seems to have been essentially constant. That is crucial for the development of advanced life. Since there are different forces that could have caused the axis direction to change, it appears to have been the stabilizing effect of our exceptionally large moon that has kept the spin axis in a stable angle to provide a consistent climate favorable to life.

How did Earth end up with a perfect 23.5-degree tilt? Is there any scientific evidence that it was ever any different? How did this tilt get coordinated in such a way that the timing when the dominant land masses in the northern hemisphere coincides with the time when the earth is farthest from the sun? Could someone or chance have made things any better?

16.

Plate tectonics

A planet with no mountains, no storms and no earthquakes will create a planet of weak people!

Mehmet Murat Ildan

Earth's surface is made up of many pieces of solid rock floating on a pliable ball of other rock that is floating on an ocean of liquid rock that has a solid steel-like center. This is a little unnerving, especially when the floating pieces of rock on the surface run into each other and shake the people living on them.

These massive collisions are called earthquakes. Also, along the edges or through holes in the floating pieces, some of the ocean of liquid rock sometimes spews out into the atmosphere and over the land masses. These holes are called volcanoes. This system seems to us laymen like a pretty scary way to build a world.

However, many researchers have stated that this system of plate tectonics (a planet whose shell is broken up into plates that constantly move around) is vital for a world to be able to host life.

Rare Earth proponents argue that the system of plate tectonics is essential for the emergence and sustenance of complex life. Drs. Ward & Brownlee assert that biodiversity, global temperature regulation, the carbon cycle and the magnetic field of the earth are all necessary for complex terrestrial life and that all depend on plate tectonics.

The system of plate tectonics is dependent on chemical composition and a long-lasting source of heat in the form of radioactive decay occurring deep in the planet's interior. Continents must also be made up of less dense felsic rocks that "float" on denser underlying mafic rock.

Earth is unique among the planets in our Solar System in that its crust is broken up into floating chunks called "plates" and these "plates" jostle about because of the convection motion of the mantle below the crust. Among other things that plate tectonics does while keeping our planet habitable is to play a key role in regulating the amount of carbon dioxide in the earth's atmosphere.

Through plate tectonics, the limestone and other carbonate minerals are carried to subduction zones where they are melted. The melted rock releases carbon dioxide through volcanoes. Volcanoes are the major natural way that the carbon dioxide in the ocean floor is released back into the atmosphere.

Automatic temperature regulation happens because of a *feedback* process that cools the earth if it gets too hot and warms the earth if it gets too cold. The rate that carbon dioxide is removed from the atmosphere depends on the temperature: the higher the temperature, the higher the rate that carbon dioxide is removed. If the earth warms up, there will be more evaporation and rainfall, resulting in greater removal of atmospheric carbon dioxide.

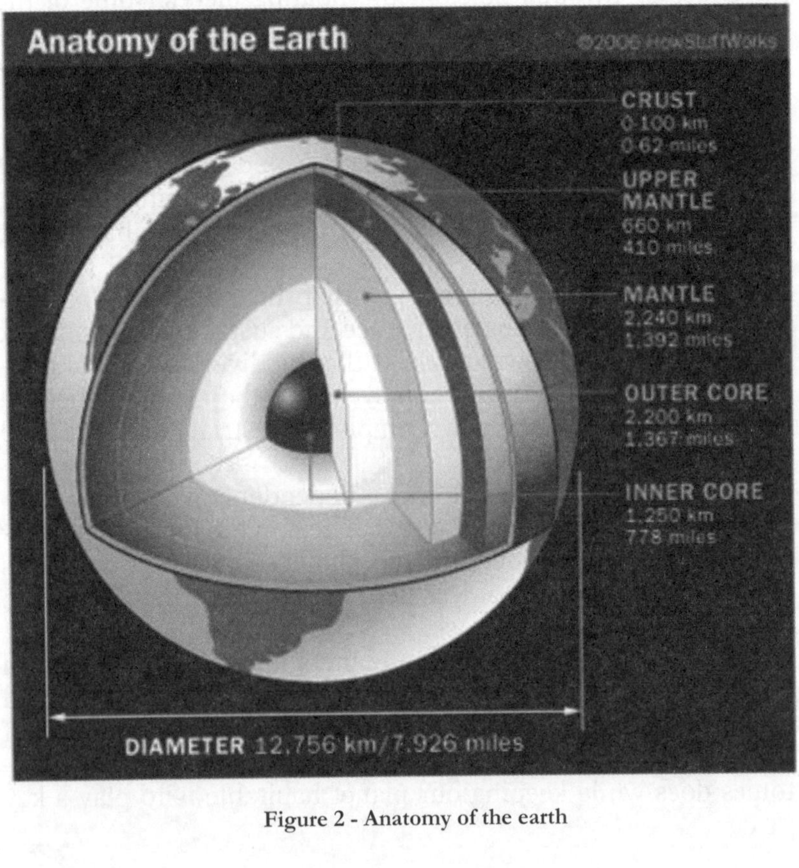

Figure 2 - Anatomy of the earth

The reduced atmospheric carbon dioxide leads to a weakened greenhouse effect that counteracts the initial warming and cools the earth back down. If the earth cools off, the rainfall decreases, resulting in less removal of atmospheric carbon dioxide.

Ward & Brownlee and others such as Tilman Spohn of the German Space Research Centre Institute of Planetary Research argue that plate tectonics provides a means of biochemical cycling that promotes complex life on Earth and that water is required to lubricate the edges of the planetary plates.

Earthquakes, even though often very destructive, are a necessary by-product of the required plate tectonic recycling. Volcanoes offer a way to vent the internal pressure. Without vents the internal pressure would build uncontrolled until the planet exploded. Volcanoes spew out the "dross" that is floating on the top of the liquid iron core of the planet and mix it with air and other gases. That's why volcanic rock is very light and porous.

Plate tectonics depends on having an active core. An active core like the earth's creates a protective electro-magnetic field all around the planet. The size of the field is important - too small and the solar winds blow the atmosphere away; too large and life is not possible.

Volcanoes are also part of the mineral recycling process. Volcanic ash covers the ground, not only providing rich soil for future generations but also in some cases creating a time vault that enables scientists to get an excellent view of the past.

To sustain plate tectonics a crust that is thick enough to support oceans and continents is required, but it can't be so thick that it doesn't have subducting plates to recycle vital minerals.

Our unique planet that is a combination of very hard solids, pliable solids, and liquids, provides for temperature control, pressure relief, recycling of carbon dioxide and minerals, mountain chains, places like Yellowstone Park, and some occasional destructive moving and shaking.

But, without this special system, Earth would have either blown apart eons ago, or would be only a dead solid rock. Could a human engineer have designed a better system?

Section III –

Our guardians in the sky

For most of human history, we have been unaware that we have guardians in the sky protecting all life on Earth from annihilation. Someone has put in place a series of protectors that have made life on our planet possible.

In recent decades, we have discovered that life outside of these protective shields is likely to be destroyed by damaging radiation. As recently as 2014, another invisible shield was discovered hiding in one of the other shields.

NASA and other teams wanting to conquer space have a great challenge because when complex life goes outside the coverage of these protectors it is in grave danger from the destructive radiation that is all around the earth, but that is not allowed to reach us.

Who knows how many more, still undiscovered, protective shields of different kinds there are making life on Earth possible?

17.

Earth's magnetic field

"Raise shields"
Captain Jean-Luc Picard in Star Trek - The Next Generation

The earth has the highest density of any of the planets in our Solar System. Its large nickel-iron core is responsible for our large magnetic field. The only other rocky planet to have any magnetic field is Mercury—but its field strength is 100 times less than the earth's. The Mariner 2 space probe found that the planet Venus does not have a magnetic field although its core iron content should be similar to that of the earth's. Venus's rotation period of 243 Earth days is far too slow to produce the dynamo effect needed to create a magnetic field around the planet.

Mars does have a very weak magnetic field, but because of its weakness, it has not served to help the planet hold onto much of its atmosphere. The Martian atmosphere contains about 95.3% carbon dioxide and 2.7% nitrogen, with the remainder a mixture of other gases. It is very thin, roughly 100 times less dense than Earth's atmosphere. Many respected scientists believe that the magnetic field is what has enabled Earth to hang on to its atmosphere until the present day.

Figure 3 - Cronodon.com

To understand the earth's magnetic field, it may help to oversimplify the situation and envision a huge bar magnet embedded in the earth that runs from the North to the South Pole. The "bar magnet" does not run exactly along the earth's rotational axis but is tilted approximately 11 degrees off the center of rotation. This skew is called the *declination*, and most good navigational maps indicate what the degree of declination is at different places on the planet's surface.

If a plane or a ship using magnetic navigation did not take into account this declination, it would not arrive at its destination. Today, with the era of the Global Positioning System that is based on satellite positioning, it is much harder to get lost in an airplane. When I was an Air Force pilot, if we did not calculate carefully the declination on a long flight, we were in danger of running out of fuel and crashing.

The "big bar magnet buried in the core" analogy works to some degree to help us understand the earth's magnetic field, but since it is impossible to have a permanent magnet above a certain temperature, there must be another explanation for what is causing the magnetism.

Permanent magnets cannot exist at the temperature found in the earth's core. When iron reaches a temperature of 770 degrees Centigrade, it loses its magnetism due to a restructuring of its molecules. Magnetism depends on the alignment of atoms in a single direction, and this alignment disintegrates in iron when this "Curie" temperature is reached.

Therefore, Earth's magnetic field must be caused by *electromagnetism*. Right at the heart of the earth is a solid inner core, two thirds the size of the moon and composed primarily of iron. At 6,000°C (10,800°F), this iron is as hot as the sun's surface, but the crushing pressure caused by gravity prevents it from becoming liquid. Surrounding this is the liquid outer core, a 1243 mile (2,000 km) thick layer of iron, nickel, and small quantities of other metals. Being at lower pressure than the inner core means the metal remains fluid.

The inner core rotates in the same direction as the earth and slightly faster, completing its once-a-day rotation about two-thirds of a second faster than the outer layers of the earth. This faster rotation of Earth's solid inner core, along with the rising of the lower parts of the liquid core due to heat, agitates and moves the molten liquid outer core. This movement generates electric currents and, as a result, creates the planet's electromagnetic field that is much more complicated than the simple field of a bar magnet.

What would happen if Earth's magnetic field became much weaker or disappeared entirely? Boy Scouts would get lost because no compasses would work. Birds, bees, sea turtles, salmon, bats, butterflies and who knows what else could not navigate and there would be no successful migrations and therefore probably the end of these species.

There would be no beautiful northern and southern lights. We would gradually lose our entire atmosphere, and most life forms would be "fried" by ultra-violet radiation from the sun.

Venus and Mars do not have magnetic fields, so they are bombarded freely with damaging radiation from the sun. Very little if any life on Earth could continue if we did not have a magnetic field.

Who made this ingenious design of a planet with a solid core, wrapped in a liquid core on which solid plates float? Why does the solid core turn slightly faster than the liquid core and faster than the hard-outer surface? What are the odds that all of these very fortuitous conditions came about by accident to create a magnetic field strong enough to protect us and yet not harm us?

18.

Who cares if Van Allen's pants fall down?

The naked truth is always better than the best-dressed lie.
Ann Landers

Earth is shielded from lethal radiation danger by five distinct invisible shields. The most prominent ones are the ones that protect us from ionized particles that come from the sun and are blown towards the earth on the solar wind, that normally takes about forty hours to arrive to Earth. The temperature above the surface of the sun is many thousands of degrees Celsius. At this temperature, collisions between gas molecules are frequent and explosive.

Free electrons and protons are thrown out from the sun's atmosphere in the form of solar flares that are caused by inner nuclear fusion and by the sun's rotation. The parts of the sun near its equator rotate once every 25 days. The Sun's rotation rate decreases with increasing latitude, so that its rotation rate is slowest near its poles. At its poles the sun rotates only once every 36 days! This is possible because it is a gas and not a solid.

Earth's first two life-saving radiation shields are produced by its magnetic field and defend the planet from deadly radiation particles coming from the sun and from outer space. This helps prevent the atmosphere from being propelled into space by collisions from cosmic particles and rays.

The magnetic field deflects most of the ionized particles as they come toward the atmosphere, but due to its weakness at the poles allows millions of particles to enter the earth's atmosphere and collide with gas particles. This is where the Aurora Borealis and the Aurora Australis come from.

What are usually called the Northern and Southern Lights are actually the result of collisions between gaseous particles in the earth's atmosphere and charged particles released from the sun. Fortunately, by the time they reach the poles, they have been deflected and weakened and can do little damage.

Because of our magnetic field, we have two layers of electromagnetic particles that protect the earth from intruders. These belts of radiation were discovered by James Van Allen in 1958. When Sputnik I, the first spacecraft, was launched in 1957 by the Soviet Union, the Americans were anxious to respond and send up their own observation satellite.

Dr. James Van Allen had already been exploring the upper atmosphere of the earth with balloons that could measure radiation levels in the atmosphere. Van Allen and his team placed a Geiger counter and an altimeter on Explorer I, the first American spacecraft, to take radiation readings at different heights.

During the flight, radiation levels increased, then suddenly dropped to zero, then increased again, and then again suddenly dropped to zero. What the team soon realized was that regions appearing as "zero" were really off the scale of the Geiger counter! These high-radiation regions were then charted and are now known as the Van Allen radiation belts.

Additional data gathered by NASA's twin Van Allen Probes, which were launched in August 2012, show that the donut-shaped Van Allen belts present a nearly impenetrable barrier to high-energy electrons, keeping them from hitting Earth.

The charged particles streaming from the sun cause the belts to grow and shrink, sometimes merging and occasionally, forming three belts instead of two. On average, the inner belt reaches from 400 to 6,000 miles (640 to 9,700 kilometers) above Earth's surface, while the outer belt spans 8,400 to 36,000 miles (13,500 to 58,000 km) above the planet.

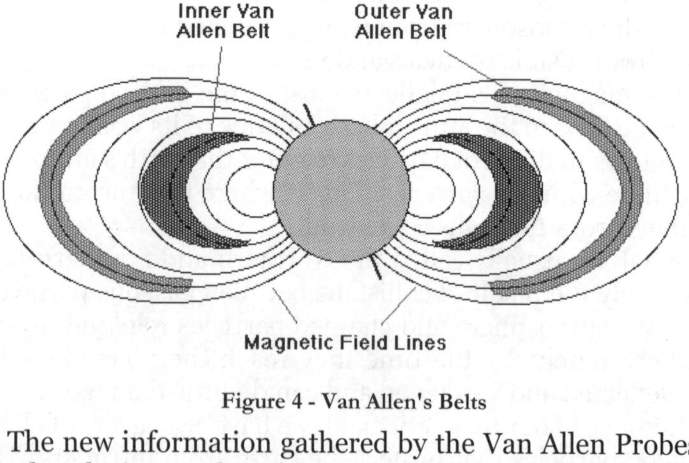

Figure 4 - Van Allen's Belts

The new information gathered by the Van Allen Probes reveals that the inner edge of the outer belt is very sharply defined.

Under normal circumstances, the fastest, highest-energy electrons cannot cross over the boundary.

"When you look at really energetic electrons, they can only come to within a certain distance from Earth," said study co-author and Van Allen Probes mission deputy scientist Shri Kanekal, of NASA's Goddard Space Flight Center in Greenbelt, Maryland. *"This is completely new. We certainly didn't expect that."*

A fundamental property of magnetic fields is that they exert forces on moving electrical charges. Thus, a magnetic field can trap charged particles such as electrons and protons as they are forced to execute a spiraling motion back and forth along the field lines.

A long-term study was done on the Apollo astronauts who traveled to the moon in the 1960s and 70s. They have suffered from a much higher rate of death by cardiovascular diseases then their fellow astronauts who did not travel outside of low Earth orbits. The researchers believe that their health problems came because they were outside the Van Allen Belts and were exposed to much higher levels of radiation.

Earth is protected by not just one, but by two invisible electromagnetic shields. If these shields were to cease to exist, so would almost all of life on Earth. Whose idea was it to make the invisible shields? If there were no Van Allen belts, Earth's "pants would fall down" and we would all be exposed to deadly radiation. Could complex life ever have developed if Van Allen's belts did not exist?

19.

A third invisible protector - the cold trap

A waffle is like a pancake with a syrup trap.
Mitch Hedberg

The temperature of Earth's atmosphere decreases with increasing altitude above the surface up to the ozone layer. On a normal day and up to about 40,000 ft altitude, it decreases about 2 degrees C or 3.5 degrees F per one thousand feet. The temperature just below the ozone layer is always below the freezing point of water.

If water vapor gets up too high in the atmosphere, it condenses and rains back to the surface. Because this height is *below* the ozone layer, the water vapor does not get high enough to be broken into hydrogen and oxygen by radiation from space. There is a "cold trap," or reflective layer below the ozone layer. This means that Earth's weather patterns almost never affect the atmosphere above 50,000 ft (9.5 miles).

If you want to avoid weather in an airplane, all you have to do is fly above this altitude (easier said than done) and the skies there are always clear. The troposphere is the lowest layer of Earth's atmosphere and the site of all weather on Earth. The troposphere contains 99% of the water vapor in the atmosphere, and the cold trap is responsible for not allowing it to move higher.

The cold trap is a protective shield in reverse that only allows water molecules to reach a certain height in the atmosphere, and then deflects them back towards Earth. This keeps hydrogen inside our atmosphere.

If the physical and chemical properties of water were not almost exactly as they are, there would be no cold trap to prevent the loss of hydrogen to space and the consequent evaporation of the oceans. As water vapor ascends in the atmosphere, it cools and condenses into liquid, forming clouds and rain and snow and falling back to the earth.

This process becomes very intense at the tropopause (the top limit of the troposphere that is 7-20 km above sea level) where air temperatures reach -80° C and all remaining water in the atmosphere is frozen.

As a result of this freezing and precipitation of water, the air above the troposphere in the stratosphere (extending up to 50 km above sea level) is absolutely dry, containing oxygen, nitrogen, some CO_2 and other atmospheric gases, but no H_2O molecules.

As air rises, its density decreases and the beneficial absorption of harmful radiation by the atmospheric gases becomes less and less effective. This means that above 80-100 km, atoms and molecules are subject to intense ionizing radiation. If water were to ascend to this level the radiation would split it into hydrogen and oxygen and, the hydrogen being very light, would be lost into space.

Over a relatively short geological period, all of Earth's water would be evaporated and the world would become uninhabitable. It is believed that Venus, having a much less effective cold trap and being so much closer to the sun, suffered this fate and lost its oceans long ago, turning into a hellish hot house.

Oxygen, which has a boiling point of -183° C, has no problems ascending through the tropopause cold trap into the stratosphere. As it does, it becomes subject to more and more intense ionizing radiation. But O_2 and O_3 as well as N_2 (Nitrogen) and CO_2 (Carbon Dioxide) are too heavy to be lost to space, and the ascent of O_2 turns some of it into O_3. This forms a protective layer in the atmosphere above the tropopause, perfectly placed just above the cold trap and preventing any ionizing radiation in the far UV range from reaching the H_2O molecules at the tropopause and in the troposphere below.

The cold trap keeps water below a certain level, and the Ozone Layer, formed by the ascension of oxygen, protects the earth from the destructive radiation from the sun and from other even worse radiation from outer space.

We amateur mechanics are amazed at the complexity of today's automobiles, and almost need a PhD in order to understand and repair them. However, the new cars are nothing compared to the complexity and beauty that keep our world functioning perfectly.

Where did the cold trap come from? Where would we be without this unique function protecting our planet? Could complex life have developed or could it continue without it?

20.

A fourth invisible shield - the ozone layer

Earth without ozone is like a house without a roof.
Anonymous

Ozone is oxygen that contains molecules that have three oxygen atoms (O_3). This molecule is tri-atomic instead of the usual O_2 molecule that is diatomic. The Ozone Layer refers to a region of Earth's stratosphere that absorbs most of the sun's ultraviolet radiation.

This layer was discovered in 1913 by the French physicists Charles Fabry and Henri Buisson. The ozone layer absorbs 97–99% of the Sun's medium-frequency ultraviolet light (from about 200 nanometers to 315 nm wavelength) that would otherwise potentially damage exposed life forms on or near the earth's surface.

Without the ozone layer, we wouldn't just sunburn, but there's a chance we'd become extinct. The sheer intensity of unadulterated ultraviolet sunlight would threaten most species that live on the planet's surface. If the ultraviolet B radiation reached Earth it would inhibit the reproductive cycle of phytoplankton, single-celled organisms such as algae that make up the bottom rung of the aquatic food chain.

Reductions in phytoplankton populations lower the populations of other animals. Researchers also have documented changes in the reproductive rates of young fish, shrimp, and crabs as well as frogs and salamanders when exposed to excess ultraviolet B.

The Ozone Layer extends from about 20 to 30 kilometers (12 to 19 mi) above Earth, though the thickness varies seasonally and geographically. It contains high concentrations of ozone relative to other parts of the atmosphere, although it is still a very small percentage compared to other gases in the stratosphere.

The ozone layer contains less than ten parts per million of ozone, while the average ozone concentration in Earth's atmosphere as a whole is only about 0.6 parts per million. It is good that this level is so low at the earth's surface, because when ozone comes into direct contact with life-forms, it displays its destructive side (hence, it is often called "bad ozone").

Since ozone reacts strongly with other molecules, high levels of ozone are toxic to living systems. This is why it can be used to kill bacteria and other unwanted germs.

The absorption of ultraviolet radiation by ozone in the upper atmosphere also creates a source of heat, that actually helps form the stratosphere (a region in which the temperature rises as one goes to higher altitudes). Ozone thus plays a key role in the temperature structure of the earth's atmosphere.

Who formed the invisible, protective ozone shield around the earth and placed it at just the right distance from us? Did it happen by chance?

Where would life on Earth be if someone hadn't made the "crazy" oxygen molecule named Ozone, and arranged for it to be in place where it does the most good and the least harm? Without it, there most probably would be no terrestrial life at all, and even aquatic life would be seriously reduced.

21.

A fifth invisible shield - the plasmasphere

First of all, never buy a man a plasma TV until he's married. A lot of men once they have a plasma TV don't need a girlfriend.
Greg Behrendt

We really don't know how many other shields might be protecting us from harmful radiation. Perhaps there are still many to be discovered. Recently, in 2014, another one was found that was not understood earlier. A team of researchers based primarily at the University of Colorado in Boulder discovered another invisible shield in space that protects Earth from "killer electrons." They comment that "it is an extremely puzzling phenomenon."

The high energy electrons that would destroy life on the surface of the earth are running into a "glass wall" located about 7200 miles above the planet. It was discovered by the twin NASA Van Allen probes that were launched in 2012 to study the Van Allen belts and how space radiation is deflected from hitting the earth.

It turns out that hidden within the Van Allen belts is a plasma shield whose origin is unknown, but that serves as an extreme barrier that allows almost no radiation to pass. This invisible shield that has been dubbed the "plasmaspheric hiss," is made up of very low-frequency electromagnetic waves that cause the space radiation to slam into neutral gas atoms in the upper atmosphere and then ultimately disappear.

It is almost as if Earth were encased in a glass bubble that prevents the bad stuff from getting in and the good stuff from getting out. It's somewhat like those old-fashioned snow globes that you turn upside down to watch the snow fall over the landscape beneath. Where did the glass bubble come from? Who knew that our fragile little planet needed a very special protector to make life possible?

22.

A sixth invisible shield - the heliosphere

When one tries to rise above Nature one is liable to fall below it. Sherlock Holmes

Not only does Earth have invisible shields, but so does the whole solar system. The solar system's deflector shield is a giant magnetic bubble called "the Heliosphere." It is caused by the sun's solar wind and magnetic field. No one knows the precise dimensions of the Heliosphere, but it's bigger than the orbit of Pluto and all of our solar system's planets are inside it.

The sun, in a different way than the earth, generates a magnetic field. The sun has no iron core working as an electric dynamo that generates the magnetic field.

Instead, it is believed that the field is caused by the dynamic movement of positively charged *ions* and negatively charged *electrons* that are bouncing around the sun in response to the nuclear reactions taking place. The sun's magnetic field fluctuates wildly and reverses polarity every 11 years. The magnetic influence propels electrons and ions at extreme distances far out into space.

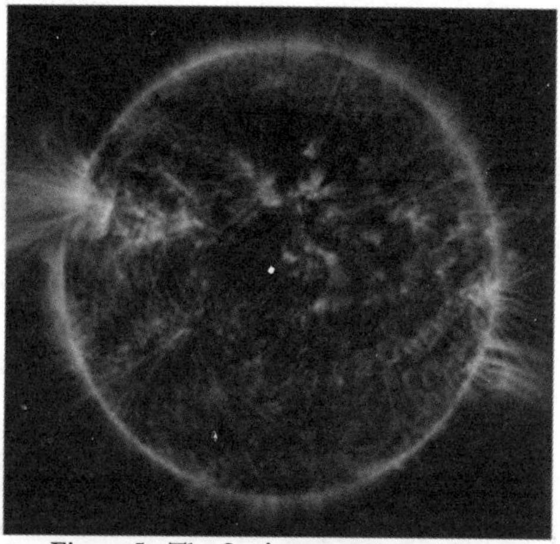

Figure 5 - The Sun's magnetic activity

The Heliosphere is important to life on our planet because it deflects much of the harmful cosmic radiation from space. This is very high energy galactic radiation that is dangerous to living things. It is believed by some astronomers that a few million years ago a cluster of massive stars drifted through our part of the Milky Way and exploded, one after another, like popcorn. Cosmic rays from the blasts were mostly deflected by the Heliosphere, sparing Earth a radiation bath that would have wiped out complex life.

Some harmful rays do get through this shield. Most cosmic rays are fragments of atoms shattered and accelerated to light speed by supernova blasts. The Heliosphere deflects about 90% of them; the rest, the most powerful 10%, penetrate to the inner solar system and to Earth's orbit where they are deflected by the electromagnetic field with its Van Allen belts, and by the plasmasphere.

The Heliosphere bubble is vulnerable to particles with no electric charge. Magnetic fields can deflect charged particles like cosmic rays, but not neutral atoms and molecules or bits of dust and rock. The bubble has an open door to these non-charged projectiles.

However, as these neutral atoms or molecules travel toward the sun, they reach a place near the orbit of the earth where sunlight knocks off an electron. When this happens, the particles become subjected to electric and magnetic forces and they are caught up in the solar wind or deflected by Earth's magnetic shields.

Perhaps astronauts should not be so eager to explore space. When they leave Earth, they get farther and farther away from their very safe environment. To survive in space, they will need to find replacements for all of the natural protections that are provided in our world. NASA, or someone, has to design space vehicles that do all of the things that our planet and solar system already do for us.

Why would they need to *design* something to imitate that which occurs "naturally?"

About six hours ago, the Heliosphere stopped a great deal of harmful space radiation that would have damaged your body beyond repair. Shouldn't you be thankful? Who installed the many protective shields around our solar system and around our planet? Did they just evolve by accident over time? Is this just another in a long line of wonderful coincidences and remarkably good luck?

23.

Don't you just love the moon?

When the moon hits your eye like a big pizza pie, that's amore." Dean Martin

From ancient times, our moon has filled mankind with wonder. It has inspired poets and musicians. For instance, an ancient Hebrew poet describes the moon as being "*firmly established for time indefinite, and as a faithful witness in the skies.*"—Psalm 89:37.

Why is the moon important? Where did it come from? One important way in which the moon affects life on Earth is that its gravitational pull causes the ebb and flow of the tides. Tidal movements are thought to be fundamental to ocean currents, which, in turn, are vital for our weather patterns and for the transport of minerals and even seeds.

Another function of the earth's natural satellite is that the moon serves as a light by night because it reflects the sun's light coming from behind the earth.

The earth has a huge Moon orbiting around it, that scientists now believe 1) did not bulge off due to the earth's high rotational speed and 2) could not have been captured by the earth's gravity (due to the moon's large mass).

The reason that our Moon doesn't fall down to the earth due to gravity is that its centripetal force that is trying to toss it into space, is balanced against gravity and therefore, although it is constantly falling, it always fails to collide with the earth.

The best explanation, other than direct creation, offered so far for the moon's existence is that a Mars-sized planet crashed into the earth around 4.25 billion years ago. The probability of two planets colliding in the same solar system is *extremely* remote. Even if one did occur, a "normal" collision would not have resulted in the formation of the moon, since the ejected matter would not have been thrown far enough from the earth to form our Moon, and would have instead created an asteroid belt leading to a constant shower of meteors.

The small planet would have had to collide with a very, very, very precise glancing blow in order to account for the angular momentum and resulting distances of the earth-Moon system. The collision of the small planet with the earth would have resulted in the ejection of five billion cubic miles of the earth's crust and mantle into orbit around the earth. This ring of material, the theory states, would have then had to coalesce to form our Moon.

In addition, the moon is moving away from the earth (currently at two inches per year – which probably coincides with the expansion rate of the universe), as it has been since its formation. If we calculate backwards, we discover that the moon must have formed just outside the Roche limit, the point at which an object would be torn apart by the earth's gravity (7,300 miles above the earth's surface).

A collision that would have ejected material to less than the Roche limit would have formed only rings around the earth. Computer models show that a collision of a small planet with the earth must have been extremely precise in order for any moon to be formed at all.

Another key purpose that our Moon serves is that its gravitational force helps to stabilize Earth's axis with respect to Earth's plane of orbit around the sun. According to the scientific journal *Nature,* without the moon the inclination of Earth's axis would wobble over long periods of time from "nearly 0 [degrees] to 85 [degrees]."

If Earth's axis had no tilt, we would miss the very important change of seasons and suffer from a shortage of rain. If it were tilted at 85 degrees, our change of seasons would be so severe that life as we know it could not survive. The earth's tilt also prevents temperatures from becoming too extreme for us to survive. Astronomer Jacques Laskar states, "*We owe our present climate stability to an exceptional event: the presence of the moon.*"

To fulfill its stabilizing role, our Moon is large—relatively much larger than the moons of the giant planets. So far, astronomers have not found another planet in the universe with such a large moon relative to its own size (the moon is $1/4^{th}$ the size of Earth). The other rocky planets in the Solar System either have no moons (Mercury and Venus), or have tiny ones that are probably captured asteroids (Mars).

The rapid rotation of our planet reduces the daily variation in temperature and makes photosynthesis possible. A planet with a large tilt will experience extreme seasonal variations in climate, unfriendly to complex life. If a person were to design a planet with just the right tilt, he could not do any better than 23.5 degrees.

If the earth did not have the moon, the ocean tides resulting solely from the sun's gravity as the earth turns would be only half that of the lunar tides. A large satellite increases the likelihood of plate tectonics through the effect of tidal forces on the planet's crust. The theoretical impact that formed the moon may also have initiated plate tectonics.

Other scientists have made intense studies of the effect that the moon has on the earth. Computer models show that the huge planet Jupiter creates a vibration in our solar system called resonance. This causes the orbits of Venus and Mercury to become highly eccentric (out-of-round). This eventually will cause a Mercury-Venus encounter that will probably result in the ejection of Mercury from the solar system, and a substantial alteration in the orbit of Venus.

In the simulations, the scientists learned that only a planet the size of Earth with a moon about the size of Mars and within 10% of the distance from the sun, would resist the resonance created by a very large planet like Jupiter. Bingo! We hit the jackpot again.

How fortunate are we that we have a Moon that serves us so perfectly? Would there be life on Earth if there were no Moon? If there were, it would certainly be different from what we have today.

24.

Jupiter is looking out for you

"The world is full of obvious things which nobody by any chance ever observes."

Sherlock Holmes, in The Hound of the Baskervilles

The four inner planets of our solar system—Mercury, Venus, Earth, and Mars—are called terrestrial or rocky planets because of their solid surfaces. The giant outer planets—Jupiter and Saturn—are composed mainly of gas. Neptune and Uranus are mostly gas, but have rocky surfaces buried very deep within their thick atmospheres.

Most astronomers believe that despite having been responsible for the shower of meteors that pelted the early Earth, the planet Jupiter and its gas planet team are now our great protectors and are responsible for collecting or ejecting a large proportion of the comets that enter into orbit around the sun.

Without Jupiter, life on Earth at this time would be difficult or impossible due to a much higher number of foreign object collisions with the earth (an increase of approximately 1,000-10,000 times more collisions created by comets, meteors and asteroids and other debris).[3]

It is surmised by many scientists that all of the dinosaurs were wiped out when a single asteroid collided with the earth near the Yucatan Peninsula in Mexico. Imagine what our Earth would look like if even 1,000 times more meteors and asteroids hit us.

Recent astronomical observation has detected a number of large planets like Jupiter in other solar systems in the universe. However, these are all too close to their stars (many orbit at a position comparable to Mercury) and therefore do not stabilize the orbits of planets in their zones, or protect any inner planets from cometary and asteroid bombardment. Instead, they contribute to destabilizing the orbits of any planets near their star.

---

[3] The Incredible design of the earth and our solar system, by Rich Deem.

The presence of Jupiter-like planets in the universe is a rare event. According to Dr. Peter D. Ward, *"All the Jupiters seen today (31 to date) are bad Jupiters. Ours is the only good one we know of. And it's got to be good, or you're thrown out into dark space or into your sun."* According to the "Rare Earth" hypothesis, without the presence of the massive gas giant Jupiter, complex life on Earth would never have arisen.

The gravitational forces of the outer giant planets Jupiter and Saturn provide the protective function of absorbing or deflecting dangerous objects headed into the inner solar system.[4]

Rare Earth proponents argue that a planetary system capable of sustaining complex life must be structured more or less like our Solar System, with small, rocky inner planets and large outer gas giants. The need for stable orbits of planets to sustain life rules out stars with systems that contain large planets with orbits close to the host star.

Another factor that makes our solar system unique is that the giant outer planets have almost circular orbits and pose no gravitational threat to the inner terrestrial planets. Imagine what would happen if either Jupiter or Saturn had elliptical orbits that brought them close to the sun. Because of the strong gravitational pull, Earth and the other small, inner planets would be thrown out of their orbits and sent on an endless journey into space, or perhaps to a fiery death in the sun.

You have to like Jupiter! Every day it is protecting you from annihilation. Complex life on Earth could not exist without this giant guardian in the sky.

---

4 *Rare Earth*, Ward and Brownlee, Copernicus , New York, 2003, 338 pages

Section IV –

Energy at work

Energy is the source of all life. Without it, objects are inanimate: hearts don't pump and brains don't think, and muscles don't move. Outer space exists because there is an on-going tug-of-war between gravity and dark energy and who knows what other forces.

Atoms exist because electrons are spinning at the right speed around protons and they are pulling on each other with the perfect amount of force to keep the electrons in orbit.

We are fascinated today with all of the *wireless* technology that has appeared in the last few decades. However, *wireless* energy has been streaming in from the sun on a constant basis to replenish the energy that we are using for our survival.

Wireless magnetic energy generated by our planet's core is protecting us from damaging radiation energy coming from outer space. Even in the dark and seemingly empty stretches of space, energy is abundant. There is no place in the universe where there is no energy. Some kinds of it are useful to us, and some kinds of it are harmful; perhaps because we don't know how to harness it or protect living organisms from it.

Before we spend our lives trying to understand how the energy works (a very valid and important vocation), we ought to find an answer to the question of where it came from.

However, let's leave that question for later, and just look at some of the remarkable forms of energy woven into the fabric of our world.

25.

Electricity

We will make electricity so cheap that only the rich will burn candles. Thomas Edison

Our modern industrial and electronic world would fall apart instantly if electricity ceased to exist. There would be no cell phones, no internet, no television, no gasoline engines, no batteries, no light bulbs, no doorbells, no cars, no computers (that operate only by a system of negative and positive charges), and no laser surgery. All but the most rudimentary industry and commerce would grind to a halt.

But it is not just our modern world that would collapse; the truth is that no animal life could ever have existed without electrical impulses.

When electrons move from one atom or molecule to another, they create an electric current. Some molecules hold their electrons very tightly and they cannot escape. Other molecules, particularly metals, hold their electrons rather loosely, and they are free to move from one atom or molecule to another. Perhaps it would help us understand electricity better if we called it "electronicity."

Electric current can consist of any moving charged particles but most commonly these are electrons. The main electricity conductor in the human body is water that makes up about 60% of the material over-all, about 73% of the brain and heart, and 83% of the lungs.

Electricity is constantly moving through the human nervous system, responding to the brain's commands in every heartbeat, muscle movement, or memory action.

Recently, on the internet, I saw a man who can handle live wires of high voltage and not get burned. The reason that he is relatively immune to electrocution is that he has a rare disease that doesn't allow his pores to sweat. Therefore, since there is no water on his skin, there is no electrical conductivity when he touches a live wire that would electrocute most people.

As mentioned, most metals have electrons that can detach from their atoms and zip around as free electrons. These loose electrons make it easy for electricity to flow through these materials, so they're known as electrical conductors. The moving electrons transmit electrical energy from one point to another.

The atoms of metal elements have valence electrons - electrons in their outer shell that are free to move about. The valence electrons are free to move through the network that forms the physical structure of a metal. When placed under an electric charge, free electrons move through the metal much like billiard balls banging against each other, and thus passing an electric charge as they move.

The most effective conducting metals are those that have a single valence electron that is free to move and causes a strong repelling reaction in other electrons. This is the case with silver and copper that each have a single valence electron that moves with little resistance and causes a strong repelling reaction. Silver has the highest conductivity of all materials, and copper is next followed by gold and aluminum.

Some of the worst conductors are Teflon, glass, hard rubber and air. They do not like to share their electrons with anyone. That's why these last items are used as resistors or insulators through which electricity finds it very difficult to pass.

It is the movement and interaction of protons and electrons, as well as neutrons and other atomic particles that cause electricity, radio waves (including light), gamma and beta rays, and other kinds of electromagnetic activity in our universe. Everything around us exists because of this interaction between positive and negative charges. If all this energy were to be taken away nothing of our world would exist.

How did our universe come to be made with just two kinds of charges – positive and negative? Why isn't there a third or fourth kind of charge? Why is all life dependent on this energy? Why is electricity necessary for the functioning of nervous systems of all complex life?

26.

Magnetism

Magnetism, as you recall from physics class, is a powerful force that causes certain items to be attracted to refrigerators.
Dave Barry

In basic science classes, nearly all students have done experiments with horseshoe magnets. We are taught that magnetism has a north pole and south pole, and that opposite poles attract and equal poles repel. Little did we know when we were playing with magnets that so many things in our life depend on magnetism.

The magnetism of the iron group of metals (iron, cobalt and nickel) is a rare and remarkable property. It is not due to any inherent magnetic propensities of the atoms, but to the structure of the metal. A property of the iron family, for which there is no substitute at an equivalent cost, is that they can be induced to provide a strong magnetic field with only a small excitation by an electric current. They can also be "trained" to have permanent magnetism like the old horseshoe magnets in science class.

What Causes Magnetism?

Magnetism is a force similar to gravity. It is caused by the movement of tiny electric charges (electrons) within atoms. When these electric charges align in the same direction, a north pole and south pole are created. North poles attract south poles, while equal poles repel each other. In natural magnets, this means that the majority of the electrons in the atoms of a magnetic element are oriented in one direction.

This causes a magnetic field that pulls or pushes on iron atoms in other particles. Any ferrous object within the magnetic field is affected by it, but some substances react more strongly. Man-made electromagnets are created by passing electrons through a wire coil that causes a magnetic field to form in the iron core.

Magnets have many applications in daily life. One of the most basic uses is the magnetic compass that is used to find geographical directions. They are also used to construct the electrical motors and the generators that convert electrical energy into mechanical energy and vice versa.

Your car has an alternator and a starter motor that use magnetism to work. They are used in Maglev trains in which the super conducting magnets are used on the tracks on which the train floats and the opposing magnetic forces lift the train. They are also used to sort out magnetic and non-magnetic substances from scrap, and in cranes to lift heavy ferrous objects. Most small gas-powered engines use magnetism to create the spark that ignites the fuel.

Magnets are used around the house in innumerable things, some of which are: kitchen appliances, antennas and radios, headphones, ear buds, stereo speakers, computer speakers, telephone receivers, phone ringers, microwave tubes, doorbell ringer solenoids, refrigerator magnets, the seal around refrigerator doors, plug-in battery eliminators, TV deflection coils, TV degaussing coils, computer monitor deflection coils, computer hard drive recording and reading heads, dishwasher and clothes washer water valve solenoids, shower curtain weights that attach to the tub, power supply transformers and many, many more uses.

Older televisions and computer monitors have magnets inside of them that make them work. The cathode ray tubes have an electron gun in the neck of the tube that shoots a stream of electrons toward the screen. Normally these electrons travel in a straight line and strike the screen at a central spot. But powerful electromagnets in the tube's neck deflect the electrons toward the top or bottom and left or right sides of the tube. The inside of the screen has a special coating that glows when the stream of electrons strikes it.

Computer storage disks are coated with an iron material that stores tiny magnetic fields in a pattern, and that is how we store data on the computer disks. VCR video tapes also used a similar material with iron compounds that allowed magnetic fields to be stored in patterns on the tape.

Floppy disks from the past and current hard drives (even in your state-of-the-art DVR) use technology that's based on the principle used by old cassette and reel to reel tapes. CD and DVD players also use magnets: instead of using magnetism to store the information, they use optical storage burned and read by lasers. However, the mechanisms used to read or create the disks use electromagnets in the motors used to drive them.

Magnets are also used to make the everyday magnetic credit cards and other forms of ID that we use. They are slowly being replaced with electronic chips that cannot be destroyed with normal levels of magnetism.

Probably the most important use of magnetism, even before people discovered compasses and electric motors, is that it is essential for your blood to carry oxygen. The iron in the red blood cell is very important because it is this that gives the blood the ability to carry oxygen around the body. These magnetic carriers store the oxygen that all parts of the body need. If the number of red cells becomes too low – in a condition called anemia – the person tends to become weaker through the lack of enough oxygen reaching the cells that need it.

Blood contains hemoglobin that is an iron-containing protein that transports oxygen around the body from the lungs to all of its other parts. In the 1930s, it was discovered that hemoglobin has magnetic properties that are different depending on whether it is carrying oxygen or not. When the hemoglobin is not carrying oxygen it is more sensitive or paramagnetic than oxygenated blood.

Without magnetism for oxygen transport in the blood, you could never have existed. Most of your machines would not run, and you would have to throw out all of your electric kitchen appliances, your cordless drills, and your motorized garden tools. Who wove the law of magnetism into the intricate fabric of the universe? Could life exist without it?

27.

Electromagnetic waves

"If you want to find the secrets of the universe, think in terms of energy, frequency and vibration."
Nikola Tesla

All forms of energy can be classified as one of three types: kinetic, potential, or electromagnetic energy (EM). Electromagnetic energy has several unique characteristics. For example, all types of electromagnetic energy can travel through empty space.

All electromagnetic energies exist as waves and travel at the speed of light. Whenever you listen to the radio, use a microwave oven, grow a tomato, tan in the sun, push a button on your remote control, or have an x-ray examination, you are using electromagnetic energy.

The type of electromagnetic energy is determined by wavelength, which is represented by the Greek letter lambda. Electromagnetic waves of different lengths possess very different properties. For instance, although visible light can pass through glass, electromagnetic waves of shorter wavelengths (UV) are blocked by glass. Also, as the length of the wave decreases, the energy of the waves increases.

Electromagnetic waves are waves that can travel anywhere in the universe, including the vacuum of outer space. Mechanical waves, unlike electromagnetic waves, require the presence of a material medium like air or water, in order to transport their energy from one location to another. The terms *light, electromagnetic waves,* and *radiation* all describe the same kind of energy. Waves are vibration and the frequency of the vibration determines what type of a wave it is.

Electromagnetic waves are similar in some ways to waves in the ocean, or to sound waves. The number of crests that pass a given point within one second is the frequency of the wave. One wave equals one cycle, and is more commonly referred to as one Hertz, named after Heinrich Hertz who established the existence of radio waves.

"EM" happens when an atomic particle, like an electron is accelerated by an electric field that causes it to move. The movement produces oscillating electric and magnetic fields that travel at right angles to each other in a bundle of light energy called a photon.

Photons travel in harmonic waves at the fastest possible speed, which is the speed of light; or perhaps more correctly stated, the speed of electromagnetic radiation.

Global positioning systems, garage door openers, wireless computer mice, keyboards, headphones, radio receivers, satellite television, broadcast television and cordless telephones all use electromagnetic waves to send information through space without physical, electrical connections.

Electromagnetic waves are in every part of the universe. These waves exist because everything in the universe is vibrating. Different kinds of waves vibrate at different frequencies. Specific boundaries do not exist between one type of electromagnetic wave and the next, but it is a continuous spectrum from one end to the other.

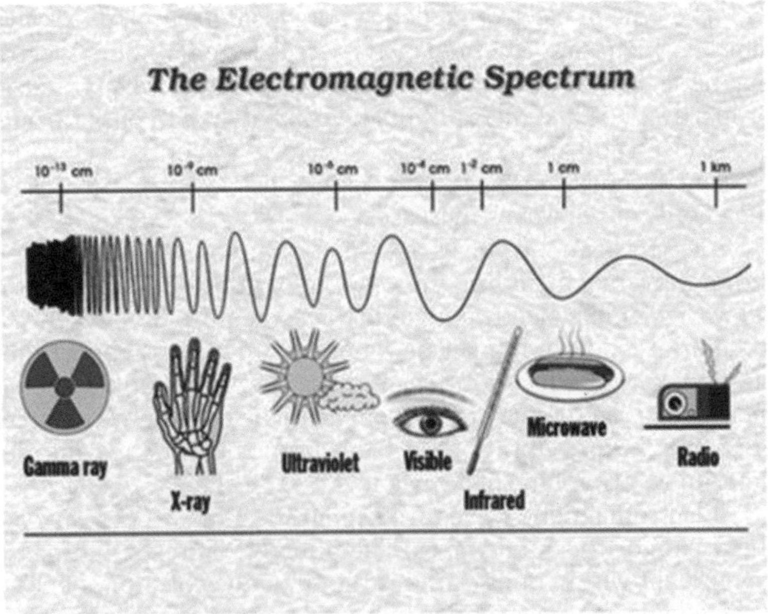

Figure 6 - https://www.teachengineering.org/lessons/view/cub_soundandlight_lesson

AM radio waves: These are the longest EM waves. Although you cannot sense radio waves, your radio can convert them into sound waves that your ear can detect.

FM radio waves/TV waves: These waves cannot be broadcast as far as AM radio waves because their shorter, more powerful wavelengths cause them to penetrate the earth's atmosphere and travel into outer space rather than being reflected back to the earth like AM waves.

Microwaves: Because water absorbs this type of energy so well, microwaves are very useful in cooking. (Living organisms, and therefore the foods we eat, are 70 - 90% water.) Microwaves (a.k.a. radar) are also used by highway police to measure the speed of cars, by civilian and military personnel to find and track flying objects, and by weathermen to find and track storms.

Infrared waves: These waves are emitted by warm objects and are also known as "heat waves." While humans can only see infrared waves if they wear a special visor, snakes can sense them and use them to hunt prey. We can feel the radiation as heat when our skin gets too close to the fire.

All of the remote controls in your home use infrared radiation to send signals, as do the fiber optics communication systems. Night vision devices, laser pointers and laser printers, CD and DVD scanners and writers, and a host of other items usually use infrared light.

Visible light: These waves are the right wavelength to stimulate the human eye's retina that sends an electrical impulse to the brain. This is the only type of EM waves that humans can "see." The longest of these waves are interpreted as being red by the human brain, the shortest are perceived as violet.

Visible light is the only type of EM energy that can be used by green plants for the process of photosynthesis. Thus, you are eating "trapped sunlight" whenever you eat food.

Visible light from the sun easily penetrates the greenhouse gases (CO_2, CH_4 etc.) of the earth's atmosphere. However, infrared (IR) waves, emanating from the warmed Earth's surface, are trapped by these gases. So, the atmosphere very "wisely" lets most of the light in, but does not allow the heat generated by it to escape into space. The more humid the air, the more it traps the heat. In a desert, night temperatures drop off much more rapidly than in a humid place.

Ultraviolet waves: These waves are energetic enough to form and break chemical bonds. A small amount of UV exposure is beneficial to humans as it stimulates the skin to produce vitamin D. Increased exposure increases the risk of severe sunburn, skin cancer and the effects of aging.

In reality, UV does stimulate the human retina and the retina does send an electrical impulse to the human brain. However, the mind does not process the electrical signal; therefore, we cannot "see" UV. Bees can see UV and flowers often reflect intricate patterns under UV. These patterns help guide bees to the nectar located in the flower's center.

Continual exposure to ultraviolet light is harmful to most organisms due to its mutation-causing effects. Fortunately, the Ozone Layer (O_3) of the earth's atmosphere absorbs most of the sun's UV rays. This radiation is often used to purify water by killing bacteria. If our bodies were exposed to large amounts of it, it would kill us.

X-rays: These waves can penetrate most substances and are commonly used as medical diagnostic tools. Like UV rays, these rays are mutagenic; they can cause mutations that damage DNA sequences.

Gamma rays: These waves are extremely energetic and dangerous to life forms. These waves are also known as cosmic waves because they emanate from outer space. There is no known upper limit of frequency.

Electromagnetic radiation is responsible for the transfer of heat and light and other forms of energy across the expanse of the universe. If there were no waves, the universe would be dead.

Where did this wave phenomenon come from? Who incorporated it into the fabric of our universe? Why do all electromagnetic waves travel at the same speed? Who set up this free system of wireless energy transfer?

28.

Light waves and vision

The only thing worse than being blind is having no vision.
Helen Keller

There are three things that can happen to a light wave. It can be transmitted, reflected, or absorbed. This is determined by the object that the wave hits, and that object will give it its color. For an object to be black it means that all the wavelengths of light hitting that object are absorbed; no light is reflected.

Solid objects, for the most part, will reflect light, and transparent objects will transmit light through them. To illustrate this last fact, place a glass of red Kool-Aid on a table. Hold a piece of white paper on one side of the glass and, if the light in the room is right, you will see red on that piece of paper. The light going through the glass is transmitting the red color of the juice onto the paper.

All light is basically made up of three colors, red, blue and green and combinations thereof. This is why we see a blue sky, green grass, and a red sunset. In our eyes, we have detectors of those three colors. If you look at old video projection devices you will see three lights made of those colors. If you examine many cables used to connect video devices, you will find those three colors.

The typical human retina contains two kinds of light cells: the 100 million rod cells (active in low light) and the 6-7 million cone cells (active in normal daylight). Normally, there are three kinds of cone cells, each containing a different pigment, and which are activated when the pigments absorb light. The frequency sensitivities of the cones differ; one is most sensitive to short wavelengths, one to medium wavelengths, and the third to medium-to-long wavelengths within the visible spectrum, with their peak sensitivities in the blue, green, and yellow-green regions of the spectrum, respectively.

These receptors are known as short (S), medium (M), and long (L) wavelength cones, but are also often, somewhat erroneously, referred to as blue, green, and red cones.

The color of anything depends on the type of light sent to our eyes; light is necessary if we are to have any perception of color at all. An object is "colored," as stated above, because of the light it reflects—all other colors are absorbed into that specific object. An apple appears red because it reflects red light and absorbs all other colors.

When you shine a light through a prism, the white light is refracted at different angles, making all seven basic colors visible: red, orange, yellow, green, blue, indigo and violet. These are the same colors seen in a rainbow where the clouds serve as a prism. The colors are split because each color has a different frequency and a different amount of energy.

At one end of the spectrum, violet has a smaller wavelength and a higher frequency (meaning more energy); while at the other end red has a longer wavelength and lower frequency (meaning less energy).

When light waves with the different frequencies are combined, the result is white light. The color "white" doesn't exist on the electromagnetic spectrum. Our eyes can be tricked into seeing white light by combining the three primary colors of light: red, blue and green. When our eyes are exposed to equal amounts of red, blue and green light, the cones absorb the wavelengths from each and tell our brains we are seeing white light.

We can only observe a very, very small portion of the electro-magnetic waves that are flying all around us day after day and hour after hour. We can probably only "sense" about 1% of the electromagnetic radiation zooming through the space around us. We need *interpreters* that can sense the rest of the radiation and convert it to frequencies that we can detect, such as visible light, or sound waves.

Without a cell phone, we can't hear wireless phone messages. Without a radio, we can't hear radio programs. Without a TV, we can't pick up the local news station. Nevertheless, the space around us is a beehive of electromagnetic activity of which we are mostly ignorant. Fortunately, we can see light, we can feel heat, and we can hear sound, all of which are important for life to exist on planet Earth.

If we had built-in sensors to feel or see all of the electromagnetic radiation, our brains would be overwhelmed with a glut of information that they could not process. Our sensors are very selective in picking up only that which is essential for survival.

Who designed our receptors to detect only the really important things? Why do we have any sensors at all? Where would complex organisms be without sight?

29.

Magnetic radio imaging

The MRI has a repertoire of noises that resemble, in no particular order: a game-show buzzer for a wrong answer, urgent knocking, a modem from 1992, a grizzly-bear growl, and a man with a raspy voice. A. J. Jacobs

During a magnetic resonance imaging (MRI) scan, technicians place you inside a tube surrounded by strong magnets, that swirl around you at very high speeds. You lie in a strong magnetic field and radio-frequency waves are directed at your body. This produces detailed images of the inside of your body, without any potentially damaging radiation like those caused by X-rays, Pet scans and Cat scans.

This imaging method is being used increasingly to image fetuses during pregnancy since it is much safer. The only drawback is that if the patient has any kind of metal implants in the body, or a pacemaker, they will heat up.

Most of the human body is made up of water molecules that consist of hydrogen and oxygen atoms. At the center of each hydrogen atom there is an even smaller particle called a proton. Protons are like tiny magnets and are very sensitive to magnetic fields.

When you lie under the powerful scanner magnets (the scanner applies a magnetic field that is about as powerful as 1000 refrigerator magnets) the protons in your body line up in the same direction, similar to when a magnet pulls the needle of a compass.

Short bursts of radio waves are then sent to certain areas of the body, knocking the protons out of alignment. Once the radio waves are gone, the protons realign and in so-doing send out magnetic signals that are picked up by very sensitive receivers.

These signals provide information about the exact location of the protons in the body. They also help to distinguish between the various types of tissue in the body, because the protons in different types of tissue realign at different speeds and produce distinct signals.

In the same way that millions of pixels on a computer screen can create complex pictures, the signals from the millions of protons in the part of the body being examined are combined to create a detailed image of the organ. It is taking a picture that is not based on visible light, but rather on other electromagnetic radiation.

What is invisible or opaque to one type of radio wave (light) is visible to a different type of wave. Did you know that the protons in your body are like miniature radio stations sending out signals telling what organ they are part of? Did you know that every living organism has a system for telling what is going on inside of it?

The "tri-corders" and other sensing devices imagined on Star Trek are not much different than some of our present-day machines except that they are a lot smaller, and certainly a lot quieter.

Man's advances in medical technology in recent decades have been incredible, but they are only a result of discovering ways to detect what is already going on inside the human body. The more we uncover the complex order of any living organism, the better we can treat anything that goes wrong with it. Living organisms have mechanisms for telling what is going on inside each cell. All we need to do is find out how to hear the message and how to interpret it.

Who established that the protons in our body would realign at different rates depending on the kind of tissue? Who made them with the ability to send any signal at all?

Section V –

Ideal conditions for life

Even the most atheistic scientists admit that our planet is a very, very special place. We have everything we need to not only survive, but to be very comfortable in many places. Like Goldilocks, everything is "just right."

We have plenty of water, oxygen, nitrogen, carbon dioxide, carbon, heat, light, and natural laws that favor our continued existence. We also have many guardians in the sky who are watching over us and protecting us from many deadly threats.

Most of us are ignorant of at least 90% of our good fortune, so it is worth taking inventory of a few of the favorable conditions that we know of that someone (or blind chance?) placed at our disposal.

Chapter V

Ideal conditions for life

[illegible]

30.

Water, water, everywhere

The ultimate aim of all science is to penetrate the unknown. Do you realize we know less about the earth we live on than about the stars and the galaxies of outer space? The greatest mystery is right here, right under our feet.

Walter Reisch

Water which is too pure has no fish.

Ts'ai Ken T'an

Earth has water! Water is believed to be the most important substance necessary for life. It contains the oxygen needed for life to exist. Most other liquids contain poisonous elements. Water doesn't harm skin (like liquids containing acids or alkalis do), it is drinkable, and it allows life-providing molecules to move around easily.

Other moons in the Solar System, such as Europa, a moon of Jupiter, are believed to have oceans of water under their icy surfaces. Scientists recently announced the finding of evidence of water on Mars, meaning that the possibility of finding some form of primitive life on that planet is greatly increased, though still very unlikely.

Water is the only common substance found naturally in all three states of matter and it is essential for all life on Earth. In nature, water exists in liquid, solid, and gaseous states. It is in dynamic equilibrium between the liquid and gas states at normal temperature and pressure. At room temperature, it is a tasteless and odorless liquid, nearly colorless but with a slightly blue tint. Many substances dissolve in water and it is commonly referred to as the universal solvent. Because of this, water in nature and in use is rarely pure and some of its properties therefore vary from those of the pure substance.

The total amount of water on the earth (in all phases) is only about 0.025% of the earth's mass, but this is absolutely essential for life. Oceans cover more than 70% of the earth: they absorb heat and distribute it around the world and strongly influence our weather patterns and climates. Water distributes nutrients to plants and other organisms. No living thing consists of less than 50% water. Water is the metabolizing agent that allows plants and animals to dissolve minerals and nutrients to convert one form of energy into another.

75% of the human brain is water and 75% percent of a living tree is water. 96% of a cucumber is water. Nearly 97% of the world's surface water is salty or otherwise undrinkable. Another 2% is locked up in ice caps and glaciers. That leaves only 1% that is available for all animal and human needs.

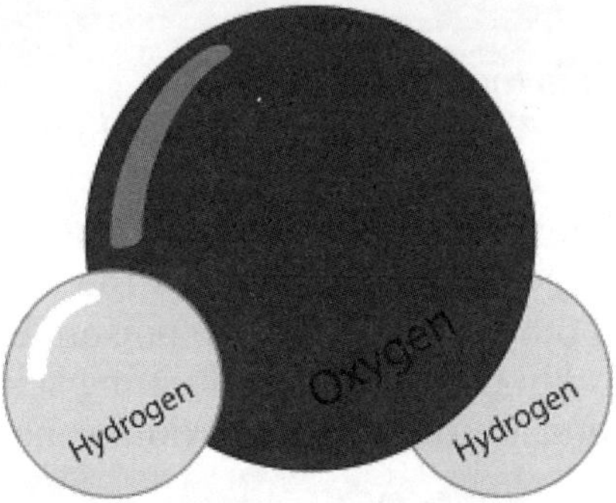

Figure 7 - Keyword-suggestions.com

An important feature of water is its bi-polar nature. The water molecule forms an angle, with hydrogen atoms at the tips and oxygen at the vertex. Oxygen has a higher electronegativity than hydrogen. Therefore, the side of the molecule with the oxygen atom has a partial negative charge compared to the hydrogen side. A molecule with this charge difference is called a dipole. The charge differences cause water molecules to be attracted to each other (the relatively positive areas being attracted to the relatively negative areas) and to other polar molecules. This is why water adheres to itself and forms droplets.

The clue to some of water's strange behavior lies with the hydrogen atoms, which each have only one electron to share with any other atom with which they combine.

When they join with oxygen to form molecules of water, each oxygen atom is balanced between two hydrogen atoms in what is called the "hydrogen bond;" but since the hydrogen atom has only one electron to offer, it can be attached firmly on only one side, so the bond is a weak one. Its strength is only about ten per cent of that of most ordinary chemical bonds. In order for water to exist, there need to be many of these bonds to hold it together.

Liquid water is so intricately laced that it is almost a continuous structure, and one scientist has gone so far as to describe a glass of water as a single molecule.[5] Ice is even more regular, and forms the most perfectly bonded hydrogen structure known.

Its crystalline pattern is so very precise that it appears to transfer over into the liquid state, and though it looks clear, water contains short-lived regions of ice crystals that form and melt many millions of times every second. If one could take a photograph with a short enough exposure, it would probably show ice-like areas even in a glass of hot water.

Liquid Water

Most of the earth's surface water is liquid. The rest is frozen or distributed in clouds in the air. Water vapor works with carbon dioxide in the atmosphere to create a greenhouse effect, raising the earth's surface temperature about 34° C. This benevolent natural greenhouse effect makes it warm enough on the surface for liquid water to exist. Besides making life possible, the liquid water also helps to keep the amount of atmospheric carbon dioxide from getting too high.

When astronomers search for life outside of our solar system, they wisely look right past the gas giants like Saturn and Jupiter, past the torrid, rocky planets like Mercury and Venus, and past the dwarf planets like Pluto.

They stop when they find an exoplanet like Gliese 581d. Gliese 581d is about 50% larger than Earth, and like Earth, it orbits in what's known as a star's habitable zone, the stellar sweet spot where a planet is capable of having liquid water. And where there's water, there might, just possibly, be life.

What makes scientists think that water is better at sustaining life than every other substance? Part of the reason is that we've never discovered an organism that's proven otherwise. While some organisms need less water than others (the

5 Some ideas taken from: http://fusionanomaly.net/water.html

cyanobacteria *Chroococcidiopsis*, for instance, needs so little water that biologists think it may be able to survive on the arid surface of Mars), every living organism we know of needs some water to survive. In fact, without water, life on Earth never could have begun.

For all simple organisms as well as the most complex plants and animals, water has played a critical role in survival since life began. In humans, it acts as both a solvent and a delivery mechanism, dissolving essential vitamins and nutrients from food and delivering them to cells.

Our bodies also use water to eliminate toxins, regulate body temperature and aid our metabolism. Water makes up nearly 60% of our bodies and we can't go for more than a few days without replenishing it.

Without water, we couldn't grow crops, keep livestock or wash our food. Water has helped spread civilization, providing a means of travel for large regions of the world and it is also a source of power for factories. Because water exists as a vapor, it is stored in the atmosphere and delivered as rain or snow across the planet. Earth's oceans help regulate the planet's climate, absorbing heat in the summer and releasing it during the winter, which moderates the average temperature dramatically. And of course, those same oceans are home for countless plants and animals.

Though it is clear that the existence of abundant water is essential to life on Earth, some have wondered if life could exist anywhere without it. Today's answer is "maybe, but not likely." Scientists are almost certain that life requires a liquid of some sort to survive, with *formamide* and *ammonia* being the most promising alternatives to water.

Both those liquids have their own set of serious problems. Liquid ammonia only exists at extremely cold temperatures, making it unlikely that organisms would have enough energy to support metabolism.

Formamide (which is actually flammable), on the other hand, stays liquid over a larger temperature range than water, and like water, it's a solvent capable of dissolving many organic materials. But so far scientists have found no evidence that it could support life.

So, we simply don't have enough information to say for certain whether or not life could exist without water. We know with certainty, however, that life on Earth or life similar to that on Earth definitely could not.

We continue to explore outer space, but so far, only 5% of Earth's oceans have been explored. Is it good science to say that God doesn't exist even though we know so little about our own planet's oceans? If we were honest, we would need to admit that we probably only know less than 1% of all there is to know about our universe.

Why is H_2O the most common molecular combination on Earth? Is it indispensable for physical life to exist anywhere? Why were we so fortunate to get so much of it?

31.

Why does water expand when it freezes?

Believing cryonics could reanimate somebody who has been frozen is like believing you can turn hamburger back into a cow.
Dr. Arthur Rowe

All liquids, solids and gases are similar in that when they are cooled (at a fixed pressure), they shrink. Objects in all three states gradually contract as they are cooled because the molecules are moving slower and are less able to overcome the attractive intermolecular forces drawing them closer to each other. In liquids, when the freezing temperature is reached, and the substance solidifies, it contracts even more because crystalline solids are usually tightly packed.

Water is one of the few exceptions to this behavior. When liquid water is cooled, it contracts like one would expect until a temperature of approximately 4 degrees Celsius (39.2 F) is reached, at which time it begins to expand. As it is cooled, it continues to expand until it reaches the freezing point and is 9% less dense. After freezing, it contracts like any other solid.

The fact that water reaches a maximum density at about 4°C causes bodies of water to freeze on the top first and also maintain a minimum under ice mean temperature of 4° C. The further expansion when it freezes keeps the ice floating with some 8% of its mass above the surface. The fact that water expands upon freezing causes icebergs, like the one that sank the Titanic, to float.

This has an interesting consequence for aquatic life in winter. Water chilled at the surface becomes denser and sinks, forming convection currents that cool the whole water body. When the temperature of the lake water reaches 4° C, water on the surface becomes less dense, and stays as a surface layer that eventually forms ice.

Since downward convection of colder water is blocked by the density change, any large body of water frozen in winter will have the bulk of its water still liquid at 4°C beneath the icy surface, allowing fish and other aquatic creatures to survive. If the water were colder than this, most life forms would perish. If, for example, the temperature in the water dropped as low as 1°C, anything still alive in it would move so slowly that you could catch it with your hand and when you did, it would already be half frozen. The temperature span of 4°C (7°F) above the freezing temperature is sufficient to keep the fish swimming and the crabs crawling until spring.

If water were most dense at the freezing point, then in winter the very cold water at the surface of lakes would sink, the lake would fill up with ice from the bottom up, and all aquatic life would be killed. Much of this ice would never thaw in the summer because the thawed water floating on top of it would insulate it.

Ocean water freezes just like freshwater, but at lower temperatures. Fresh water freezes at 32 degrees Fahrenheit but seawater freezes at about 28.4 degrees Fahrenheit, because of the salt in it.

The reason that water expands when freezing is that when it is near its freezing temp (below 4°C), the hydrogen bonds between water molecules become stronger. Just prior to being frozen and while frozen the water molecules are arranged in a manner that forms a six-sided crystal that makes the ice less dense than the liquid water.

Other substances that expand upon freezing are silicon (freezing point 1414°C / 2577°F), gallium (29°C/86°F), germanium (938°C/1720°F), antimony (630°C/1137°F), bismuth (271°C/519°F), and plutonium (639°C/1182°F). The only one of these elements that occurs in abundance on Earth is silicon, and except in a laboratory, a volcano or a glass factory, you will never see it in liquid form.

Water covers more than 70% of the earth's surface, and if it did not expand when freezing, no life could exist in rivers and lakes far from the equator, at the poles, or in most of the oceans north of the Arctic Circle and south of the Antarctic Circle. It is even questionable whether marine life could exist outside of the tropics since, each year, water that froze in the winter would continue to accumulate on the ocean bottom until the oceans were frozen solid.

Ships could only navigate in the tropics, and the important green-house effect (not usually the villain it is made out to be) would be diminished so much that our planet would gradually cool itself into death.

Where would we be if water didn't expand on freezing? Who tweaked the system to prevent the extinction of life in lakes, rivers and oceans when it turns cold? Who set the temperature under the ice at a perfect 4°C? How did such a wonderful liquid (as well as solid and gas, for that matter) come into existence in such quantities on our planet in order to provide for the winter survival of aquatic life?

32.

Carbon, the indispensable building block for life

Life exists in the universe only because the carbon atom possesses certain exceptional properties.

Sir James Jeans

Of the more than 90 known natural elements, only about two dozen can be found in most living organisms. Some of these are only found in trace amounts, and only six make up the majority of an organism's composition: carbon, hydrogen, nitrogen, oxygen, phosphorus, and sulfur.

Most biological molecules are made from covalent combinations of these six important elements, whose chemical symbols are CHNOPS. Biological molecules, or biomolecules (any molecule that is present in living organisms), are built by joining atoms through covalent bonds (chemical bonds involving the sharing of electrons between atoms in a molecule, especially the sharing of a pair of electrons by two adjacent atoms).

Although more than twenty-five different elements can be found in biomolecules, the six elements mentioned are most common. They comprise 98-99% (dry weight) of all living organisms.

Carbon forms more compounds than any other element. Because of its willingness to bond to other nonmetallic elements it is often referred to as the building block of life. Carbon is the 4th most common element in the universe (after hydrogen, helium and oxygen). It is only the 15th most common element in the earth's crust but it is the second most common element in the human body.

Carbon has the highest melting point of all elements, around 3500° C (3773° K, 6332° F). Today's plastics are made from carbon polymers and carbon is used to form alloys with iron such as carbon steel. It is used in a wide range of applications like carbonated beverages that employ carbon dioxide gas.

Life based upon carbon requires that water exist in the liquid state (within a very narrow range of 100°C). For all practical purposes, this range is even narrower. Researchers have discovered only a very few bacteria that can exist near the boiling point of water, but they are very specialized. Nearly all known life forms must exist below a temperature of 50°C (122°F).

Carbon forms the key component of all known physical life. It is the basis for all plants and animals, but also diamonds, charcoal, petroleum, plastics and graphite. Plants extract atmospheric carbon dioxide in the photosynthesis process and use it to form organic compounds. Most of that carbon dioxide is released back into the atmosphere when plants decay or are burned.

Much organic material gets deposited in marine sediments under the ocean. If the temperature and pressure conditions underground are right, the organic material in the marine sediments may be converted into oil, natural gas or coal. This holds the carbon in storage until these organic-based fuels are extracted and burned. Burning them releases their carbon back into the atmosphere in the form of carbon dioxide.

Aquatic plants extract carbon dioxide dissolved in the water to use in their photosynthesis process. Aquatic animals use the carbon dioxide and calcium in the water to make shells of calcium carbonate ($CaCO_3$). When the animals die, their shells settle to the ocean floor where, after years of compacting and coalescing, they lock up the carbon dioxide when forming limestone.

The majority of the earth's carbon dioxide is buried deep below the ocean's surface in the form of carbonates.

Carbon is also the basic building block required to form proteins, carbohydrates and fats, and it plays a crucial role in regulating the physiology of the body. Carbon (18%) is second only to oxygen (65%) in the composition of the human body. Hydrogen makes up 10%, and that leaves only 7% for the rest of the elements.

Life on Earth, which to date is the only type of physical life we have discovered, is exclusively carbon based. Life depends on the favorable conditions on Earth, including its abundance of water, its oxygen-rich atmosphere, and its climatic stability, to prosper. But what makes carbon so special? Are there also other types of life possible that are not based on carbon?

The possibility of alternate life forms has been the topic of investigation for numerous scientists in recent years. They have been considering silicon as a possible alternative basic building block for life, but so far that research has come up empty.

Recently, some NASA researchers have found a different kind of bacterium that uses arsenic in place of phosphorus, but that bacterium still cannot exist without carbon.

Later research discovered that even this exceptional bacterium went to great lengths to prefer phosphate instead of arsenate, even when "trapped" in a medium with 500 times as much arsenic as phosphorus. Like all other known organisms, it was doing its best to avoid the deadly arsenic, but was able to survive despite its presence, not because of it.

So far, there is no known physical life apart from the element carbon. Carbon also has many other uses. Without it, school kids wouldn't have any pencils, brides wouldn't have diamond rings, oil-well drillers and dentists would have to drill longer (ouch!), there wouldn't be fuel for cars, boats, airplanes or fireplaces, or petroleum-based electricity generating plants. There would be no plastics, no bubbly in sodas, no cookouts on the 4th of July, and no burnt toast.

Could our universe exist without carbon? If it did, it is very unlikely that there would be any life in it. Where did this indispensable element for life come from? Why is carbon so abundant on our planet? What if it were as scarce as Gold?

33.

All complex life depends on oxygen

Simplicity is the ultimate sophistication.
Leonardo Da Vinci

Our atmosphere is crucial to any kind of surface life. It consists of a layer of gases about 50-60 miles thick that is very thin compared to the size of Earth (3,959-mile radius or about 8000 in diameter). That means that our atmosphere is just a little over 1% as thick as the earth.

When complex living organisms go more than 5 miles above the earth's surface (.15% of the earth's radius), unless they carry their oxygen with them, they can no longer survive.

As mentioned, the atmosphere also helps maintain a relatively consistent range of temperatures and shields the earth from collisions with cosmic objects both large and small (a falling or shooting "star" is one of these large particles being burnt up by the atmosphere).

The atmosphere is made up of gases like nitrogen 78.09%, oxygen 20.95%, argon 0.93%, carbon dioxide .039%, and small amounts of other gases. The oxygen level of 21% allows for life; if it were to drop very far, complex animal life would not be possible; if it were to exceed 25% most plants would be consumed by fire.

The combination of gases in the air has a molecular weight of 1.29 g/l. Hydrogen, which is 14 times less dense than air, has a weight of .090 g/l, and helium has a weight of .178 g/l. This means that both of these gases will rise in the atmosphere because they are lighter than air. Oxygen weighs 1.41 g/l and therefore descends in stable air.

Air is mostly nitrogen and the density of nitrogen gas is 1.24, making it the main determiner of the mass and weight of air. If the density of oxygen were lower than that of air, all of the oxygen in the atmosphere would have long since passed to the higher environs of the atmosphere and there wouldn't be any here on the surface for animals to breath. Nitrogen, the other gas indispensable for the growth of plants has a tendency to rise, but only very slightly.

Oxygen is produced by plants and is consumed by animals and plants. The greenhouse effect, produced by carbon dioxide and water vapor, allows for heat in the atmosphere to be retained by re-radiating it back to Earth.

The crucial difference between plants and the rest of living organisms is that plants are the net 'givers' of energy and all the other living organisms are 'takers.' Almost all organisms need green plants for control of CO2 and the manufacture of oxygen and the provision of food at the base of the food chain.

The exceptions are the strange creatures living at deep sea heat vents. Those creatures depend for food on tiny bacteria that get their oxygen from the water around them through a process called chemosynthesis.

Earth has a breathable atmosphere due to a number of complex factors. Oxygen is the gas that is required for the existence of all known animal life. This is present in Earth's atmosphere and also in water. Oxygen is constantly put into the atmosphere by plants.

Earth's atmosphere also contains a very small amount of carbon dioxide that is a bi-product of animal breathing, of burning, and of decaying processes. This is the gas that makes up most of the atmosphere of planets like Venus and Mars and makes them unable to support human life.

The "just-right" smaller percentage of CO2 on Earth is necessary for the growth of plants that use it for photosynthesis and produce oxygen as a bi-product. It also helps moderate the planet's temperature.

Earth's atmosphere is kept on the planet by the pull of gravity. Mars, Mercury and Pluto are too small to retain significant atmosphere. As a result, Mercury has no atmosphere, and Mars' atmosphere is very thin, only containing gases that have not managed to escape into space yet.

If oxygen, nitrogen, carbon dioxide or hydrogen were missing from our atmosphere, or if the percentages were changed by a small amount, there would be no complex life on our planet.

Why is Earth the only planet in the solar system (and perhaps anywhere) to have high enough amounts of oxygen in its atmosphere to sustain life? Who arranged for the delicate balance of these elements so that life could exist?

34.

Please pass the salt

No man is worth his salt who is not ready at all times to risk his well-being, to risk his body, to risk his life, in a great cause.
Theodore Roosevelt

If salt disappeared from the earth, the wheels of industry would eventually stop and it would be only a matter of months before we'd all be dead.

As mentioned earlier, our planet, Earth, is the only confirmed place in our solar system to have liquid water. Recently, NASA has said they have discovered evidence of water on Mars, but that is still unconfirmed by solid (or liquid) evidence.

Astronauts looking at Earth's surface from high in our atmosphere see mainly water. The ocean covers 71% of the Earth's total surface area, and contains enough water to cover the whole planet to a depth of 1.7 miles (2.7 km) if the earth's surface were completely flat (as well as round).

The ocean is essential for life on Earth, and also helps make the climate fairly moderate. Although the ocean contains 1,370 million cubic kilometers (334 million cubic miles) of water, humans can't survive by drinking from it—it is too salty.

To a chemist, 'salt' refers to a wide range of chemicals where a metal is combined with a non-metal. Ordinary table salt is a compound formed when the metal sodium combines with the non-metal chlorine to form sodium chloride. This contains electrically charged atoms called *ions* that strongly attract each other, resulting in a fairly hard crystal. When salt dissolves, these ions separate. Sodium and chloride ions are the main ions in seawater, but not the only ones. The salty seas benefit man because the ocean provides many useful minerals for our industries and food for our bodies.

We could not live without salt. An article in *This Week Magazine (*newspaper insert that ran from 1935-1969) reported that exceptionally strong men, working in the forge rooms of an automobile factory, where the temperature was so high that the men perspired abnormally, would drop to the floor and writhe in agony without apparent cause.

After prolonged study, Harvard Laboratory staff discovered that, although the men drank cold water freely due to the hot temperature, this did not replace the salt that they had perspired out of their systems. Taking a salt tablet with each drink of water stopped the "heat cramps."

While in college, I worked for two summers in a cabinet manufacturing plant. My job was to work at the delivery end of a long drying oven that had conveyer belts carrying the cabinets from the spray-painters out to the finishing line. It was never less than 95 degrees there, and in the afternoons got closer to 110. Needless to say, we were sweating and drinking lots of water.

The company must have read the *This Week* article because it had wisely placed two salt tablet dispensers on the walls near the ovens, and we usually took a few of those each day to replace the salt lost by perspiration.

Not only do our bodies need salt, but so does the manufacturing world. Steel, subjected to terrific strain in automobiles, is hardened by a salt water bath. The famous Carrel-Dakin antiseptic solution that saved hundreds of thousands of lives during World War I and is still in use today, has as its chief ingredient chlorine that comes from salt.

Salt has been historically indispensable to support life. In Britain, the suffix "wich" in a place name means it was once a source of salt, as in "Sandwich" and "Norwich." The word "salary" comes from the Latin word for salt because the Roman Legions were sometimes paid in salt. They say the soldiers who did their job well were "worth their salt." The word "salad" also originated from "salt," and began with the early Romans salting their leafy greens and vegetables.

The famed Erie Canal, opened in 1825, was known as "the ditch that salt built" because salt, a heavy and bulky product presenting major transportation difficulties, was its most important cargo. In 1777, the British Lord Howe was jubilant when he succeeded in capturing General Washington's salt supply. Thousands of Napoleon's troops died during their retreat from Moscow because their wounds would not heal due to the lack of salt.

Without salt, cows stop giving milk, sicken and die. Farmers provide salt supplements and salt blocks for their cattle, and hunters like to put them in the forest to attract deer and other wild game because all animals need salt.

A few reasons why salt is important:

- Our bodies rely on electrolytes (that come from dissolving salt in water) to help carry electrical impulses that control many of our bodies' functions. To keep our bodies functioning as they should, our bodies need the proper amount of various kinds of electrolytes. Electrolytes trigger the thirst mechanism that causes us to consume adequate amounts of water. With this water, our kidneys are able to keep the appropriate amount of electrolytes in our bloodstream. The amount of water our bodies retain also impacts blood pressure.
- Salt helps to relieve muscle cramps.
- Salt helps retain water in the body.
- Iodized table salt and "natural" sea salt contain iodine that is important because it keeps you from getting hypothyroidism and a goiter in your thyroid gland (Don't confuse "natural" sea salt with the super white, de-mineralized crystalline variety that you buy at the grocery store).
- Salt plays a primary role in the processes of digestion and absorption. Salt activates an enzyme in the mouth called salivary amylase that allows your taste buds to taste your food. Salt also plays a role in digestion by helping to break down food. Salt also helps create hydrochloric acid, which is a very important digestive secretion that lines the stomach walls. Salt actually helps your body digest food rather than trying to eat itself.
- Salt is important to the nerves, as it helps control muscle contractions, thus preventing your muscles from cramping. Salt keeps calcium and other minerals in the bloodstream. It also stimulates the adrenal glands.
- It is indispensable in the prevention of heat prostration and sunstroke, which is very important to remember during hot summer months.
- Sodium deficiency can lead to shock if the blood pressure is decreased too severely.

• Salt is important to good nutritional status. Too little can cause disturbances in tissue-fluids and in the acid-base balance.

• All our body fluids: blood, sweat, tears, saliva, etc. are salty.

• The brain and spine rest in a sac of salty liquid called cerebrospinal fluid. This liquid also circulates throughout the brain and spinal column.

• We all spend about 9 months floating in a sea of salt liquid (amniotic fluid) in our mother's womb.

• Our bones are hollow in the center where blood cells are made. The marrow at the center is covered with many strands of calcium salts, like an intertwined rope. The salt crystals in this covering help make our bones hard.

• Brain signals are transmitted to the muscles and other organs through ions produced by salt.

• Electrolytes, which are a product of various salts dissolved in water, carry glucose into the cells after insulin opens the access door.

• A proper level of salt in the body helps reduce bladder control problems and urine leakage, and also helps bring down blood sugar levels.

• A proper supply of salt helps to prevent osteoporosis, varicose veins, and helps maintain an erection.

• Salt may even help reduce a double chin. When low on salt, the salivary glands in our mouth produce more saliva and this often leaks into the lower part of the jaw.

• Rock salt is important for melting ice on roads so that you can survive on winter days while riding in your vehicle.

• Salt was used before refrigeration to keep meat and other food from spoiling. Salt, in a process known as osmosis, draws water out of the cells of both the food and the micro-organisms that want to destroy it. Micro-organisms that do not have enough access to water will not grow and reproduce as fast as those with ample access to it. Salt helps to prevent spoilage of the food by reducing its access to fresh water.

- Salt also inhibits bacteria by disrupting enzymes in microbes and damaging their DNA.
- Salt therefore can be used to heal wounds by reducing infections.

Our oceans have an abundance of salt. Most scientists believe that the saltiness is increasing, though no one in recent years has come up with clear data to prove or disprove that claim. We do know that there is a constant flow of new salt going into the oceans. Rain and ground water dissolve salt deposits, especially in clays and feldspars, and wash the sodium out of them and this salty water is then carried to the ocean.

Some salt is supplied by water through the ground directly to the sea—called submarine groundwater discharge (SGWD). This water is usually very concentrated in minerals. Sediments on the ocean floor as well as hot springs (hydrothermal vents) release sodium. Volcanic dust also contributes some sodium.

Without the salt compounds, complex life on Earth could not survive. Why is there such an abundance of this vital material in our world? Who set up the rules for the formation of salty compounds? Who made salt an indispensable ingredient for nearly all, if not all, living things?

35.

Photosynthesis

When you think about the complexity of our natural world - plants using quantum mechanics for photosynthesis, for example - a smart phone begins to look like a pretty dumb object. Jeff Vander Meer

By using the energy of sunlight, plants can convert carbon dioxide and water into carbohydrates and oxygen in a process called photosynthesis. Since photosynthesis requires sunlight, this chemical reaction only happens naturally during the day, though it can be accomplished at night by means of artificial lighting.

We often like to think of this process as plants `breathing' in carbon dioxide and `breathing' out oxygen. However, the process is much more complicated than that. Just like animals, plants need to break down carbohydrates into energy, and oxygen is required to do this.

The good news is that plants do not need all of the oxygen they produce during photosynthesis. Plants do use a small amount of the oxygen they produce to break down carbohydrates to give them energy to live and grow.

What happens at night when there is no sunlight for photosynthesis? For plants to maintain their metabolism and continue respiration at night, they must absorb oxygen from the air and give off carbon dioxide (which is what animals also do).

Fortunately, plants produce approximately ten times more oxygen during the day than what they consume at night, so 90% of it becomes available to oxygen-hungry animals.

So, all green or vascular plants consume oxygen all the time. They release a lot of oxygen through photosynthesis but they consume some of it at the same time. The primary way plants use oxygen is through their root cells. In order for a plant to take up water and nutrients from the soil, it must continually grow new root hairs.

The cells that make up these hairs are living cells that do not carry out photosynthesis. They are similar to the cells of animals because they require oxygen and give off carbon dioxide as they burn carbohydrates to obtain the energy they need to function and grow. You can "drown" a plant by watering it too much.

If the soil gets too saturated with water, the plants root cells can't get enough oxygen, unless they happen to be vascular plants such as cypress trees and cattails that can live in saturated soil. These plants have root hairs that grow out of the stems above the soil or water surface so that they can take in oxygen. Have you ever seen the cypress roots sticking up out of the ground or the water in Florida or Louisiana?[6]

Photosynthesis is the process by which light energy is converted into chemical energy. Using the energy of light, plants manufacture carbohydrates, such as sugars, from carbon dioxide and the hydrogen in water. The name photosynthesis is derived from the Greek words, *photo* for 'light' and *synthesis* meaning 'putting together.' After the plants take hydrogen from the water, they release oxygen as a waste product.

Glucose, a carbohydrate produced during photosynthesis, is used by plants as construction material to build leaves, flowers, fruits, and seeds. Molecules of glucose combine with each other to form more complex carbohydrates such as starch and cellulose. The cellulose is the structural material used in plant cell walls. Photosynthesis provides the basic energy source for virtually all living organisms because the plant carbohydrates, proteins and fats become food for animal and parasitic plants.

Where does Photosynthesis occur? –

Photosynthesis takes place primarily in leaves, and almost none occurs in stems. It takes place within specialized cell structures called chloroplasts. A leaf has a stalk and a lamina, the flat portion of the leaf. Because this area is broad, the lamina helps in the absorption of sunlight and carbon dioxide during photosynthesis.

Photosynthesis takes place in the chloroplasts which have chlorophyll present in them. It is the chlorophyll that absorbs light energy from the sun. There are tiny pores called stomata

---

[6] Some ideas taken from: https://answers.yahoo.com/question/index?qid=20060622214812AACGseV

that function as channels for carbon dioxide to enter and oxygen to leave the plant.

When a plant is dormant and its leaves have fallen off as in the case of perennial deciduous plants, the plants are still using oxygen and giving off carbon dioxide through their roots since the roots must stay active to preserve the root crown from which the new stems and leaves will emerge during the next growing season.

Compared to other known planets, the earth has an unusual atmosphere! The presence of free oxygen (O_2) is very unusual and necessary for plant life. Oxygen just loves to chemically react with other atoms and molecules. The oxygen in our atmosphere would soon disappear (within about 5000 years) if photosynthesizing organisms like plants and cyanobacteria did not regenerate the oxygen supply.

Since plants keep oxygen in the form of O_2 and its fragile cousin, ozone (O_3) around, to find evidence of more than just a smidgen of these two molecules in the atmosphere of a planet beyond our solar system would be one reason to expect life to be possible on that planet. So far, no such planet with more than just a trace of gaseous oxygen has been discovered.[7]

Why plants are green -

Did you know that plants only use a part of the spectrum of visible light? The chlorophyll absorbs and uses blue, red and violet light rays. The light that is absorbed the best is blue after which comes red light. Green light cannot be absorbed by the plant, and thus cannot be used for photosynthesis.

Chlorophyll looks green because it absorbs red and blue light, making these colors unavailable to be seen by our eyes. It is the green light that is not absorbed that is therefore reflected and reaches our eyes, making the chlorophyll and therefore plants appear green.

The rate of photosynthesis depends on the intensity of the light source, the concentration of carbon dioxide, the presence of

---

7 http://curious.astro.cornell.edu/about-us/63-our-solar-system/planets-and-dwarf-planets/comparisons-with-earth/263-is-there-oxygen-in-the-atmospheres-of-other-planets-intermediate

sufficient water, and the temperature. It can be slowed by pollutants in the atmosphere that restrict the arrival of light, or by the scarcity of water.

How would you like it if trees and plants were red or blue instead of green? Or worse yet, if plants absorbed all three basic colors and all you saw was black? Where did the idea for chlorophyll come from? Who thought of the extremely complex process of photosynthesis?

Did the process of photosynthesis evolve from nothing by successive throws of the dice? Where would plants and animals be without this fundamental pillar of life?

36.

Wind

"The pessimist complains about the wind; the optimist expects the wind to change; the realist adjusts the sails."
William Arthur Ward

It is estimated that 1-2% of the sun's energy on Earth is converted naturally into wind energy. There are several purposes for which wind exists. Wind is here to help cool the earth, the trees, animals, and humans, and also to distribute water to most places. We would not be able to live without wind.

There is wind because the earth is revolving around its axis, and because of uneven heating of the oceans and the land. If the earth did not revolve, there would be no significant wind and therefore no distribution of heat or moisture across the surface of the planet.

Wind is mainly air in motion relative to the earth's surface. We normally think of the wind as the horizontal motion of the air, although air really moves in three dimensions. The vertical component of the wind is generally quite small, except in thunderstorm and tornado updrafts.

The vertical motion of air, however, is very important in determining our weather. Air that is rising cools, which causes it to reach its saturation point, form clouds and condense in precipitation. On the other hand, air that is sinking warms, which causes clouds to evaporate and produce clear weather.

Wind can be thought of as one way that the atmosphere distributes excess heat. Wind forms for the primary purpose of helping to transport excess heat either away from the surface of the earth, where sunlight causes an excess of energy buildup, or from warm regions (usually the tropics) to cooler regions.

Surface weather maps usually have "High's and "Low's at various locations. The H's and L's represent high- and low-pressure systems. On weather maps, places with the same atmospheric pressure readings are connected by lines called *isobars*.

When isobars are close together, the pressure is changing rapidly over a small distance. The closer the isobars are to each other, the stronger the *pressure gradient* (the rate of pressure change over a given distance.) In the Northern Hemisphere the wind blows clockwise around a high-pressure system and also slightly outward from its center.

Around a low-pressure system, the wind blows counterclockwise and slightly in towards its center. In the Southern Hemisphere, the direction of rotation is reversed (we observe that the drawing shown above is from the southern hemisphere because the wind arrows are going in a counter-clockwise direction around the high-pressure center and clockwise around the low).

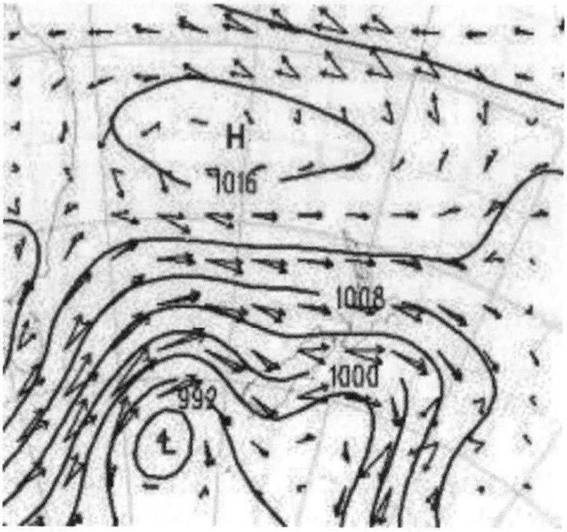

Figure 8 - A typical surface weather map

Why does the wind blow? There are three forces that cause the wind to blow in the direction that it does:

1. Pressure Gradient Force
2. Coriolis force (caused by the earth's rotation)
3. Friction

Pressure Gradient Force –

The *Pressure Gradient Force* (PGF) is due to differences in pressure between one area and another. This force is trying to move the air to equalize pressure differences. The PGF causes air to flow from high pressure to low pressure. The greater the pressure difference, the higher the wind velocity.

When looking on a surface map, strong winds occur in locations where the isobars are very close together. This pressure gradient is also what makes a vacuum cleaner work because an electric fan creates a low-pressure area and air rushes to fill it.

Coriolis force -

This is an inertial force that causes fluids (like water and air) to curve as they travel across or above the earth's rotating surface. The *Coriolis force* exists because of the earth's rotation. This force causes moving objects (i.e. air, airplanes, birds, artillery shells, etc.) to deflect to the right of their motion in the Northern Hemisphere and to the left in the Southern Hemisphere. It is this Coriolis force that causes large storms like hurricanes, cyclones and tornados to spin around their central axis.

If the earth were not spinning, then the wind would occur in nearly a straight line (Example - A newspaper boy on a bicycle knows that he has to throw the paper before he gets to the place he wants it to land because the paper will travel almost the same distance forward as he does before it hits the ground – or sometimes a bush or the house roof).

Friction -

The third force acting on the wind is friction. Friction becomes very important near the earth's surface because the surface of the earth is not completely smooth. Friction is the force that causes air to slow down and spiral into lows and out of highs.

During the day, when convective mixing due to sunlight is stirring up the lower atmosphere, the influence of friction becomes less significant. At night when the heating effect of sunlight is gone, the surface wind can slow considerably, or even stop. Wind velocity often drops to near zero when the sun goes down.

Wind is also important for the transfer of pollen from one plant to another, and for the distribution of many seeds to fertile ground. It also accounts for the spread of insects.

Wind plays a major role in a tree's life. Resisting the wind makes a tree stronger so it is able to mature and not fall down due to its own weight. Experiments show that if there's no wind, the trees end up being much weaker and aren't able to survive for long.

Where would life be without wind? Some places on Earth would be even more humid than they are and others even more dry. Others would be even hotter and others colder. There would be few or no moderate or temperate climates where crops could grow well. We would have either muggy swamps or arid deserts, and very little in between.

No ships could have sailed the seven seas, no kites would have ever flown, there would be no windmills to visit in Holland, and no breeze would have ever blown to cool your sweaty brow.

How did the earth figure out it needed wind to support life? Was this another case of the survival of the fittest, or the design of an extremely wise Creator?

37.

Evaporation and condensation

He decided how hard the winds should blow and how much rain should fall. Job

Evaporation is the process by which a liquid or a solid changes into a vapor (gas). A substance may evaporate in one of two ways: (1) by changing into a vapor at the surface, as when water evaporates from an uncovered dish; or (2) by boiling, that is, by changing to vapor both within the substance as well as at its surface.

A solid can evaporate by melting into a liquid, which then evaporates; or by changing directly into a vapor (sublimating). The rate of evaporation of a substance depends on its molecular structure, its surface area, its surface temperature, the atmospheric pressure, the concentration of the substance already in the air, and the humidity.

Evaporation is caused by heat that is the energy of molecules in motion. Molecules are motionless at absolute zero (-459.67 F/-273.15 C.). Above that temperature they are in constant motion and the higher the temperature the faster they move. In moving, the molecules bump into each other. With each collision they transfer energy.

If two or more molecules bump a surface molecule in quick succession, it may gain enough energy to escape into the air. An evaporating molecule takes energy with it. Loss of this energy cools the substance from which it has escaped.

There are endless uses for evaporation. It cools our bodies as our perspiration converts from liquid to gas. The evaporation of water from the land into the atmosphere keeps the air moist and makes clouds that provide rain and snow to remoisten the earth. Approximately 80% of all evaporation of water is from the oceans, and the remaining 20% comes from inland water and vegetation.

Winds transport the evaporated water around the globe, influencing the humidity of the air throughout the world. For example, a typical hot and humid summer day in the Midwestern

United States is caused by winds blowing tropical oceanic air northward from the Gulf of Mexico.

Evaporation is the process used for air conditioning, refrigeration, and distilling. Instant coffee and frozen orange juice are evaporated in a vacuum, as are blood plasma, vaccines, and antibiotics such as penicillin.

Evaporated water (steam) was used for many years as a propellant for locomotives, ships, industrial machines and even cars. There would have been no industrial revolution without steam power. Today, steam is still used in most power stations (nuclear, coal, oil and gas fired) to spin turbines that generate electricity.

Who made evaporation and condensation, without which there would be no rain and therefore no crops growing on most of the planet's surface? No one would live in an air-conditioned house or drive an air-conditioned car. There would have been no steam heating of houses, hospitals and schools for hundreds of years, and there would be no old steam trains or paddle-wheelers to see in museums or tourist destinations.

Without evaporation, there would be no distilled spirits, either for drinking or medicinal purposes.

Were we just lucky that when our world was formed, this law of nature was thrown into the mix? Could life on our planet exist without evaporation and condensation? Would I live in Florida without air conditioning?

38.

Capillary action

"Look at those cows and remember that the greatest scientists in the world have never discovered how to make grass into milk."
Michael Pupin

Capillary action is defined as the ability of some liquids to flow in narrow spaces without the assistance of, and in opposition to, external forces like gravity. It causes many liquids to be drawn up or lifted; in a thin tube, between the hairs of a paintbrush, in porous materials such as paper, in some non-porous materials such as liquefied carbon fiber, or in a cell. It is the result of intermolecular forces between the liquid and surrounding solid surfaces.

Capillary action happens because water is "sticky" due to the forces of cohesion (water molecules stick together) and adhesion (water molecules stick to other substances). The adhesion of water to the walls of a vessel will cause an upward force on the liquid at the edges and result in a concaved upper surface (lower in the middle).

The surface tension of the molecules holds the surface intact. Capillary action results when the adhesion to the walls is stronger than the cohesive forces between the liquid molecules. The height to which capillary action will take water in a uniform circular tube is limited by surface tension and gravity: the narrower the tube, the higher the movement.

Not only does water tend to stick together in a drop, it sticks to glass, cloth, organic tissues, soil, and, luckily, to the fibers in a paper towel. Dip a paper towel into a glass of water and the water will "climb" into the paper towel. In fact, it will keep going up the towel until the pull of gravity is too much for it to overcome. If you lay a paper towel over the edge of a bowl of water, the water will climb up and then start flowing down the other side. In fact, it will flow down the other side a lot faster because now gravity and capillary action are pulling in the same direction.

The first recorded observation of capillary action was by Leonardo Da Vinci in the 15th Century. In 1660, capillary action was still a novelty to the Irish chemist Robert Boyle, when he reported that "some inquisitive French men" had observed that when a capillary tube was dipped into water, the water would ascend to "some height in the pipe."

Even though you may never have heard of capillary action, it is still very important to you. Capillary action affects all kinds of things in our daily life:

- It is essential for the drainage of constantly produced tear fluid from the eye. Two canaliculi (extremely narrow tubular channels) are present in the inner corner of the eyelid, also called the lacrimal ducts. When these get clogged up, glaucoma can result.
- Wicking is the absorption of a liquid by a material in the manner of a candle wick. The small pores of a sponge act as small capillaries, causing it to absorb a large amount of fluid. Some textile fabrics use capillary action to "wick" sweat away from the skin. These are often referred to as wicking fabrics, after the capillary properties of candle and lamp wicks.
- Kerosene lamps are good examples of capillary action at work in the wicks to transport the fuel to the flame. When a dry porous medium, such as a brick or a wick, is brought into contact with a liquid, it will absorb the liquid at a rate that decreases over time.
- Capillary action draws ink to the tips of fountain pen nibs from a reservoir or cartridge inside the pen, and even up into the inner chamber of a feather allowing quill pens to function. John Hancock relied on capillary action to sign the Declaration of Independence.
- In hydrology, capillary action describes the attraction of water molecules to soil particles. Capillary action is responsible for moving groundwater from wet areas of the soil to dry areas. Without capillary action, water under the ground would stay at the water table and never rise to the surface to meet the plants. We are told to cultivate our gardens and fields to break up the big cracks that cannot support capillary action and to recreate finer soil that can.
- Trees and other plants couldn't thrive without capillary action. Plants put down roots into the soil in order

to carry water from the soil up into the plant. Water that contains dissolved nutrients gets inside the roots and starts climbing up the plant tissue.

You can see capillary action by doing an experiment where you place the bottom of a celery stalk in a glass of water with food coloring and watch for the movement of the color to the top leaves of the celery. You might want to use a piece of celery that has withered because it is in need of a quick drink.

It can take a few days, but the colored water is "drawn" upward, against the pull of gravity. This effect happens because, in plants, water molecules move through narrow tubes that are called capillaries (or xylem).

Although our heart pumps our blood and pressurizes it, there is not enough pressure to cause it to pass through all of the really small channels in our bodies. As the arteries and veins get thinner and thinner, the pressurization has less and less effect on the flow.

It is capillary action that makes sure blood gets to even the remotest part of the body to reach all of the body's seventy trillion cells. Of the 50,000 miles of the blood circulation system in the body, capillaries comprise over 49,000 miles.

If capillary action did not exist, you would not exist nor would any other mammals or birds or fish, because all of these creatures that circulate blood depend on capillary action to survive.

Who included capillary action in the design of our world? Did it just happen? Where would life be without it?

39.

Buoyancy

Up, up and away

Superman, in the old comics and radio series

Buoyancy keeps everything from ducks to ocean liners afloat. Without this common force that exists in all liquids and gases, nothing would swim on the surface of water. Divers wouldn't need weights to go under water, but they would need a crane to get them up from the bottom. Submarines would need to be all-terrain vehicles and crawl along the ocean floor.

Ice wouldn't float in your glass of coke, and boat manufacturers would all go out of business. The only ships that could travel in the ocean would be ones that had wings like airplanes and could only sustain themselves as long as they had forward motion. If they stopped moving, they would crash to the bottom.

Archimedes discovered the law of buoyancy about 250 BC. He allegedly yelled out "Eureka!" ("I found it") and ran naked through the city of Syracuse. He knew that the discovery was that important. The story goes that Archimedes made his great breakthrough when he noticed the water rise as he got into a public bath.

Buoyancy, first and foremost, is a force like gravity. Buoyancy, however, moves a body in an upward direction always in the opposite direction of gravity. This upward force occurs only in a fluid. This can be a liquid or a gas, mediums that have moving electrons less bound to each other than in solids. It cannot be a solid, because there the electrons are held firmly in place.

According to Archimedes' buoyancy principle, the force acting on, or buoying, a submerged or partially submerged object equals the weight of the liquid that the object displaces. This principle has an immense range of applications and is essential to calculations of density, as well as designing submarines and other oceangoing vessels.

Archimedes' Buoyancy Principle -

Any object, wholly or partially immersed in a fluid, is buoyed up by a force equal to the weight of the fluid displaced by the object.

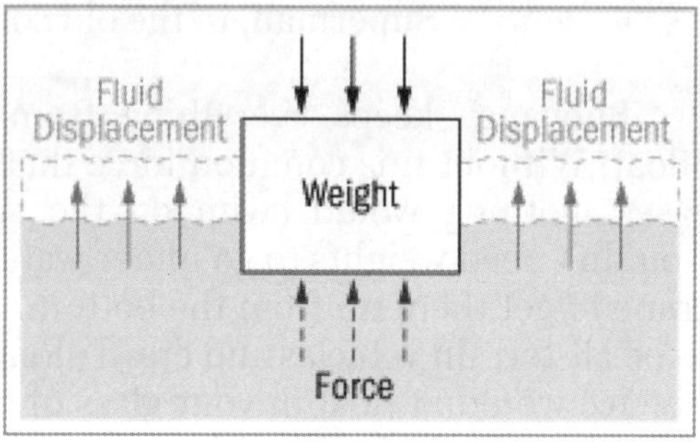

Figure 9 -
www.google.com/search?q=archimedes+principle

The principle applies to both floating and submerged bodies and to all fluids. It explains not only the buoyancy of ships and all kinds of other vessels in water but also the rise of a balloon in the air and the apparent loss of weight of objects going under water.

In determining whether a given body will float in a given fluid, both weight and volume must be considered. If the object is less dense than the fluid, it will float or, in the case of a balloon in the air, it will rise. If the object is denser than the fluid, it will sink.

Relative density also determines the degree to which a floating object will be submerged in a fluid. If it is two-thirds as dense as the fluid, then two-thirds of its total volume will be submerged and it will displace (push out) a volume of fluid whose weight is equal to the entire weight of the object.

The fluid most often encountered in applications of Archimedes' principle is water (because it is the most common liquid). In calculating the buoyant force on an object, one must also take into account the shape and position of the body.

An aluminum rowboat placed on end into the water will sink because the density of aluminum is greater than that of water. However, in its keel-down position, the effective volume of the boat includes all the air inside it, so that its average density is less than that of water, and as a result the boat will float.

Fish have a swim bladder that manipulates neutral buoyancy by controlling the amount of air in the chamber, allowing them to swim at varying depths. A fish's density becomes higher or lower than the surrounding water due to its changing density based on the amount of air in its bladder. Submarines have a similar system of tanks with alternating air or water. When the tanks are full of water, the submarine sinks: when they are full of air, it floats.

The functioning of the human brain depends on neutral buoyancy, which it attains as a result of its suspension in cerebrospinal fluid (CSF). The actual mass of the human brain is about 1400 grams but the net weight of the brain suspended in the CSF is equivalent to a mass of 25 grams. The brain therefore exists in near neutral buoyancy that allows it to maintain its density without being impaired by its own weight that would cut off the blood supply and kill neurons in the lower sections.

Without the law of buoyancy, you couldn't take your kids to the local pool, have egg-drop soup, or use a bobber while fishing. Dross couldn't be scooped off the top of molten silver or gold, there would be no buoys in the ocean, no dirigibles at football games, and no kids crying because they lost their helium birthday balloons. There would have been no iceberg to sink the Titanic, but neither would there have been a ship by that name.

Eric the Red would have never discovered America (before Columbus), we would never have heard of Marco Polo, and they would have to eliminate all the water sports at the Summer Olympics.

If life could have moved forward at all, it would have been restricted to the land where it began. And then also, if our brains had gotten damaged by the constriction of its blood vessels, we wouldn't be thinking about where the world came from. Who thought to program our world with this indispensable ingredient named buoyancy?

40.

Bernoulli's Principle - why things fly

Oh! I have slipped the surly bonds of Earth and danced the skies on laughter-silvered wings; . . . put out my hand, and touched the face of God.

John Gillespie Magee

Have you noticed the curved shape of an airplane's wing? A bird's wing is also curved. There are curved airfoils on airplanes, helicopters, propellers, turbines and many fans. These items take advantage of Bernoulli's Principle that states that when the speed of airflow over an object is increased, the pressure over them decreases.

Since the top surface of an airplane's wing has more camber (curve) than the bottom surface, the air has to flow faster over the top of the wing than it does underneath. This means that there is less air pressure above the wing than there is beneath the wing, and the difference in air pressure above and below causes lift.

A flat plate held at an angle of attack in relation to wind direction generates lift, but it also generates a lot of drag. Sir George Cayley and Otto Lilienthal during the 1800's showed that curved surfaces generate more lift and less drag than flat surfaces. Early research also showed that a round leading edge and a sharp, flat trailing edge add to a wing's ability to generate more lift and less drag. Airfoils are patterned somewhat after the shape of a whale and other sea creatures.

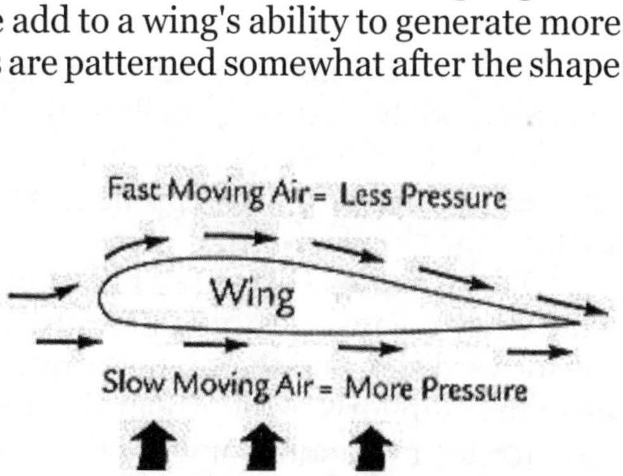

Figure 10 - www.livescience.com

Daniel Bernoulli (1700 – 1782) was a Dutch-Swiss mathematician and was one of the many prominent mathematicians in the Bernoulli family. Bernoulli's principle states that as the speed of a moving fluid (liquid or gas) increases, the pressure within the fluid decreases.

The phenomenon described by Bernoulli's principle has many practical applications; it is employed in the carburetor and the atomizer, in which air is the moving fluid, and also in the aspirator, in which water is the moving fluid.

In the first two devices air moving through a tube passes through a constriction that causes an increase in speed and a corresponding reduction in pressure. As a result, liquid is forced up into the air stream (through a narrow tube that leads from the body of the liquid to the constriction) by the greater atmospheric pressure on the surface of the liquid.

In the aspirator, air is drawn into a stream of water as the water flows through a constriction (This happens when you turn on the bubbly feature of many Jacuzzis or hot tubs). Bernoulli's principle can be explained in terms of the law of conservation of energy (the First Law of Thermo-dynamics says that energy can be neither created nor destroyed).

As a fluid moves from a wider pipe into a narrower pipe or a constriction, a corresponding volume must move a greater distance forward in the narrower pipe and thus have a greater speed. Since the speed is greater in the narrower pipe, the kinetic energy of that volume is greater. This increase in kinetic energy must be balanced by a decrease in the pressure.

Birds "came" with curved airfoils for wings. Though there are different shapes for different types of birds, they all work on the principle that as air flows over them, they create a difference in pressure that makes the wing "fly" upwards.

If there were no Bernoulli's Principle, neither birds, nor airplanes could fly. Instead of making the trip on a modern jet from London to New York in eight hours, you would need to go by ship, except that ships have propellers that also use airfoil shapes to drive the ship forward.

Old-fashioned ships used sails, but when the sails filled with wind, they created a curve that caused the air to go faster over the forward side and therefore the difference in pressure helped push the ship forward. Without this Principle, you would likely have to stay in London, or cross the ocean in a stern-wheeler, or be shot out of a cannon.

Was the Bernoulli Principle here when the world began, or did it evolve from nothing? If it was here when the world began, what stroke of luck put it here?

41.

Honeycomb structure efficiency

Oh honeycomb, won't you be my baby? Oh honeycomb, be my own."

From the classic song *Honeycomb* by Jimmie Rodgers, 1957

The most efficient and strong structure that man has ever designed already existed in nature before man discovered it. In 1859 Charles Darwin stated that *the comb of the hive-bee is absolutely perfect in economizing labor and wax.*

Hexagonal honeycomb patterns often appear in nature, where strength, rigidity and lightness are called for. The shells of armadillos, the beaks of birds and, of course, wax cells built by bees are just a few examples of nature's honeycombs. Engineers have long known about the honeycomb's strength and low density, and the structure has been used in applications as varied as satellite components and the "scaffolding" for growing new heart tissue.

Figure 11 - biomimicrysandiego.org/biomimicry-education

Honeycomb structures permit manufacturers to build materials with maximum strength while keeping the volume and therefore the cost of the materials low. Man-made honeycomb structural materials are commonly made by fastening a honeycomb material between two thin layers of material that provide tensile strength. This forms a sandwich-like assembly. They are widely used in the aerospace industry because they are both lightweight and strong. That's why honeycomb materials in aluminum, fiberglass and advanced composite materials have been featured in aircraft and rockets since the 1950s.

They can also be found in many other fields, from packaging materials in the form of paper-based honeycomb cardboard, to sporting goods like skis and snowboards. Internal doors of houses are often just two layers of thin plywood, glued to a cardboard honeycomb interior. Aramid fiber honeycomb is also used in aerospace applications, but often in more advanced systems, such as missile and hull components.

This type of honeycomb is also used in various marine and military structures. Polypropylene honeycomb is used in noise and vibration reduction applications, such as acoustical panels.

Honeycomb shaped containers hold more liquid and can be stacked next to each other with no lost space in between. Square containers also leave no space in between, but their walls must be thicker to achieve the needed strength. Circular containers are very strong, but waste a lot of space.

Who taught the bees how to make honeycombs, or even how to make honey? Why can't engineers and scientists find a better structure than the one that was already here?

42.

Sound and hearing

When the whole world is silent, even one voice becomes powerful. *Malala Yousafzai*

Sound is the result of vibrations in the molecules of a medium (solid, liquid or gas) causing the molecules to compress and then expand to make a longitudinal wave. The vibrating source must "push" some material (i.e. air, metal, water) for the sound wave to travel. Therefore, sound cannot travel in a vacuum or a near vacuum like outer space where there is almost nothing to carry the wave.

All material objects can vibrate to make a sound. All objects will naturally vibrate at a specific frequency known as their *natural frequency.* A big thick guitar string vibrates naturally at a low frequency. A smaller, thin guitar string vibrates at a high frequency.

In order to hear sound, we must have a sound sensor – an instrument that can detect the movements coming through the medium. This might be compared to having a radio that converts radio waves into sound waves. Our human ear can only detect sounds with frequencies between 20 Hz and 20,000 Hz, and with amplitude (strength) strong enough to be within our ear's sensitivity range.

The eardrum is a very thin membrane that stretches across the ear canal at the threshold between the outer ear and middle ear. When sound waves hit the eardrum, the eardrum vibrates and sends the vibrations to the middle ear, where they pass through the hammer, anvil, and stirrup (This is the smallest bone in the human body - it is 0.25 to 0.33 cm long and weighs .0001 ounces).

These three tiny bones remain the same size from birth to death. If their relation varied the least bit, any sound would be distorted or not heard at all. The amount of movement at the tip of the stirrup is less than the diameter of a hydrogen atom, but it is sufficient to communicate the difference between the drop of a pin, the buzz of a mosquito, and thunder.

The stirrup bone passes the vibrations along the cochlea in the inner ear. Inside the cochlear duct is a strip of tissue known as the organ of Corti, one of the most remarkable organs in the body. This organ essentially consists of three rows of outer hair cells and one row of inner hair cells.

These hairs are actually cilia that are much smaller than a skin hair on our bodies. In fact, they are too small to be seen individually, even with a light microscope. The wiggling of the hairs causes small molecular "trapdoors" on the tips of the hairs to open and close, permitting electrically charged particles (ions) to enter the hairs. Incredibly, the molecular trapdoors are controlled by molecular springs that attach to adjustable molecular brackets.

It is amazing to understand that tiny molecular trapdoors are opening and closing at a rate between 20 and 20,000 times per second, admitting charged ions into the tips of the hairs. This movement of ions generates electrical signals that are sent to the brain, where they are processed and interpreted.

Sound is a disturbance passing from molecule to molecule, like a string of falling dominos. The speed at which the molecules in a medium can move determines the speed limit in that medium. Air, which is the most common medium for sound waves, has a built-in speed limit of about 344 meters per second (1238 km/h or 770 mph) at sea level and at average temperature. If something goes through the air faster than that, it breaks the air sound barrier and creates a sonic boom.

This speed limit in each substance varies depending on the density of the medium. The speed limit in water is five times faster than in air. Ask someone to stick their head under water and then tell you if they hear it when you bang two rocks together under the water. Of course, they will have to take their head out of the water in order to say anything you could hear.

The speed limit in steel is 17 times faster than air. The density of a substance varies according to temperature. The lower the density of the medium, the more temperature affects the speed.

When a human's sound sensor (ear) begins to fail, he is often given a hearing aid to amplify the signal. If the sensor fails completely, they can now do what is called a cochlear implant that is an electronic sensor implanted on the outside of the skull and attached to the inner ear that stimulates the auditory nerve.

A cochlear implant is very different from a hearing aid. Cochlear implants bypass damaged portions of the ear and directly stimulate the nerve. Signals generated by the implant are sent by way of the auditory nerve to the brain, which then recognizes the signals as sound. Hearing through a cochlear implant is very inferior to normal hearing and takes time to learn.

Hearing is the universal alerting sense in all vertebrates (creatures with a bony spine). Sound is very important because animals are able to hear events all around them, no matter where their attention is focused. Many species of blind amphibians, reptiles, fishes and mammals are known, but no naturally profoundly deaf vertebrate species has been discovered.

Although hearing is important to all animals, the special qualities of the undersea world emphasize the use of sound. Sound travels far greater distances than light under water. Light travels only a few hundred meters in the ocean before it is absorbed or scattered. Even where light is available, it is more difficult to see as far under water as in air, limiting vision in the marine environment.

The undersea world presents very different conditions for hearing as well as seeing. The sounds produced by many marine mammals can project for miles. Strong echoes are always present underwater, because sound travels without much loss and there are many underwater surfaces that reflect sound.

Marine mammal sounds are probably structured so that they can be recognized in spite of all the echoes. Marine creatures can perceive sound coming from much further distances than terrestrial animals or birds. Because the sound travels faster, they also receive the sounds after much shorter delays.

Marine mammals, such as whales, use sound to identify objects such as food, obstacles, and other whales. By emitting clicks, or short pulses of sound, marine mammals can "listen" for echoes and detect prey, or navigate around objects. This animal sense functions just like the sonar systems on navy ships. It is clear that producing and hearing sound is vital to marine mammal survival.

Sound is also important to fish. They produce various sounds, including grunts, croaks, clicks, and snaps, that are used to attract mates as well as ward off predators.

Why did whales do podcasts long before it ever occurred to humans? Why does every material in our universe vibrate at its own particular frequency when stimulated? Who made the speed limits for each material? Who enforces the speed limits and keeps them constant?

43.

Touch

And I sometimes think that a moment of touching is the difference between complete utter despair and the ability to carry on." Eleanor Cameron

Without our sense of touch, it would be very hard to get around in this world! We wouldn't feel our feet hitting the floor when we walked, we wouldn't sense when something sharp cut us, and we wouldn't feel the warm sun on our skin. It is truly amazing how much information we receive about the world through our sense of touch, and although we still don't know all the ins and outs of how the skin perceives touch, what we do know is fascinating.

Touch "senses" when our "glob of energy," called our skin, comes in contact with some other "glob of energy," that we perceive as a physical object. Our sense of touch is controlled by a huge network of nerve endings and touch sensors in the skin known as the *somatosensory system*. This system is responsible for all the sensations we feel - cold, hot, smooth, rough, tickle, itch, pressure, pain, vibrations, and more.

Within the somatosensory system, there are four main types of receptors: mechanoreceptors, thermo receptors, pain receptors, and proprioceptors.

When drinking from a glass of soda, your hand can perceive many different sensations just by holding it. Thermo receptors are sensing that the glass is colder than the surrounding air, while the mechanoreceptors in your fingers feel the smoothness of the glass and possibly the small fluttering sensations caused by the carbon dioxide bubbles rising to the surface. Mechanoreceptors located deeper in your hand sense that your hand is around the glass and that pressure is being exerted to hold it.

Proprioceptors are also sensing the hand stretching out as well as how the hand and fingers are holding the glass in relation to each other and the rest of the body. Your somatosensory system is probably sending lots more information to the brain than what was just described.

The neurons in the nervous system of the body take up the important task of relaying messages from the skin to the brain. Neurons come in many shapes and sizes. Their shapes and connections help them carry out specialized functions, such as storing memories or controlling muscles.

When your hand touches an object, the mechanoreceptors in the skin are activated, and they start a string of messages by signaling to the nearest neuron that they touched something. This neuron then transmits this message to the next neuron which gets passed on to the next neuron and on it goes until the message arrives at the brain.

Information from one neuron flows to another neuron across a small gap called a synapse (SIN-aps or sin-APS). At the synapse, electrical signals are translated into chemical signals in order to cross the gap. Once on the other side, the signal becomes electrical again. Any given neuron will have about a thousand synapses with neighboring neurons, connecting the cells and allowing them to send messages from neuron to neuron. The synapses in a single human brain outnumber the stars in the Milky Way (10^{11}).

The brain can process what your hand touched and send messages back to your hand via this same pathway to let the hand know if the brain wants more information about the object it is touching or if the hand should stop touching it. This two-way communication usually takes about .1 seconds.

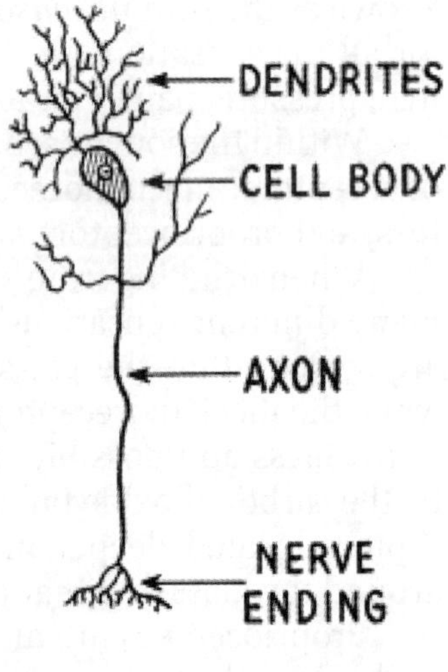

Figure 12 – prozacmonoogues blogspot.com/2011 06/ getting-my-brain-back-in-praise-of-bdnf.html

Like all animal cells, a neuron has a cell body, called a soma, where the DNA-carrying nucleus resides. This gives directions for the cell to make various proteins. In a neuron, however, this is just the beginning of the cell structure. On one end, the soma sprouts branch-like *dendrites* for receiving signals, while a long *axon* stretches away in the other direction, branching out into multiple *axon* terminals for sending signals.

A protective covering called the *myelin sheath* insulates the axon and helps the electrical nerve signals travel faster and farther.

These axon terminals are located close to the dendrites of other neurons — though the axon terminals do not physically touch the other neuron's dendrites. The synapses are empty space with no direct connection between one neuron's axon terminals and another's dendrites.

The axons disperse a mixture of different chemicals and send them across the gap. Within each axon terminal are sacs, known as vesicles, filled with one of 50 different chemicals called neurotransmitters. Each neurotransmitter sends a different type of message to the next neuron that recognizes the neurotransmitters with specialized coded receptors on the surface of the dendrites.

Although chemicals are required to send a message from neuron to neuron, the signal is converted back to electricity to transmit that message from the receiving neuron's dendrites to its own axon terminals. When the neurotransmitters trigger the receiving neuron to fire, it sends an electrical pulse along its length the way that an electrical current flows down a metal wire.

These electrical pulses are caused by ions building up within the cell. As mentioned earlier, these ions are products of salts dissolved in the body.

Of the five senses (touch, taste, smell, sight, hearing), touch probably functions in the most similar way for animals on land as well as underwater. For underwater animals, the sense of touch may also provide important information about water currents and the motions of nearby animals.

Today's fancy cars have over one hundred different electrical components, and it has taken tens of thousands of engineers, and over a hundred years to get to this degree of sophistication. How could a complex electrical system like our mind, containing more than 10 billion cells and more connections than stars in our galaxy, evolve without any supernatural help?

If the chemical mixtures in the neurotransmitters are off just a little, the brain would tell the hand to slap the face instead of taking another drink of tea. Isn't it absurd to believe that such a delicate, complex system could make itself with no outside help?

Section VI –

Who wrote the laws for our universe?

Our world has many laws that govern how things work. There is a pattern and a purpose to the universe and to all life forms. Here are a few examples of laws, some of which are not well known:

1. The Law of Gravity – mass attracts other mass at a constant rate.
2. The speed of electromagnetic radiation (including light) is unchangeable.
3. The total charge of the positively charged particles in the universe shall be equal to the total charge of the negatively charged particles.
4. The mass ratio of protons to electrons can never change greatly – experiments prove that this ratio can have changed only by a maximum of one part in a million over the age of our Solar System.

The same laws of nature we find on earth also govern a star billions of light years away. You can act in accordance with these laws, or you can try to disregard them, but you can in no way alter them.

Where did the laws come from? Who wove these laws into the fabric of our universe and who enforces the laws? Could anything exist if there were no laws?

44.

The Second Law of Thermodynamics

Nothing in life is certain except death, taxes and the second law of thermodynamics. Seth Lloyd

Thermodynamics: *the study of heat power; a branch of physics that studies the efficiency of energy transfer and exchange.*

The Second Law of Thermodynamics states that as energy is transferred or transformed, more and more of it becomes unavailable for conversion into mechanical work. The Second Law also states that there is a natural tendency of any isolated system to degenerate into a more disordered or random state.

This law is based on the fact that the useable energy in the universe is decreasing. The universe is running down, and ultimately there will be no useful energy left. As a result of this fact, we see that the most probable state for any natural system is to move from a more orderly state to a less orderly one. All natural systems, without input from outside sources, degenerate into chaos when left to follow their own course.

This Second Law is a law of universal decay; an explanation for why everything ultimately falls apart and disintegrates over time. Material things are not eternal. Everything left to itself deteriorates and chaos increases. This law applies to everything in the universe.

Nothing stays as fresh as the day one buys it; clothing becomes faded and ultimately returns to dust. Everything ages and wears out. The decay and death of all living things, including our bodies, is probably the most sobering consequence of this law. The effects of the Second Law are all around us, touching everything in the universe.

Each year, vast quantities of time, effort and money are spent to counteract the relentless effects of this law (maintenance, repairs, painting, medical bills, etc.). Everything in nature is subject and obedient to its unrelenting influence.

Through the use of relatively large amounts of energy, outside forces can increase order in a system for a time. However, such intentional ordering will not last forever. Once the outside force is removed, processes return to their natural state – increasing disorder. Chemical compounds over time break apart into simpler materials; they do not naturally become more complex.

There is an overall downward trend throughout the universe. Ultimately, when the useful energy of the cosmos has been depleted, all molecules will move more slowly and more randomly, and the entire universe will be cold and without order.

Finally, everything will reach a point where no molecules can move because there is no usable energy left in the system. All experimental and physical observation confirms that this Second Law is indeed universal, affecting all natural processes in the long run.

Many scientists believe that the Second Law, when truly understood, is enough to refute Darwin's Theory of Evolution that says that elements organized themselves into molecules, then into amino acids, then into proteins and then into living organisms that became increasingly complex.

The Theory of Evolution goes 180 degrees in the opposite direction of this Law that has never been disproved.

This basic law of science reveals that elements in a closed system do not organize themselves. No experimental evidence disproves it, say physicists G.N. Hatspoulous and E.P. Gyftopoulos:

"There is no recorded experiment in the history of science that contradicts the second law or its corollaries..."

If all of the evidence shows that things do not make themselves, but rather fall apart, who or what put our world together and created the extremely complex structures of living organisms? Who or what organized our world into such an orderly place? Science says that it couldn't have happened on its own. What do you say?

45.

Newton's Laws

The motions which the planets now have could not spring from any natural cause alone, but were impressed by an intelligent Agent. Isaac Newton

Sir Isaac Newton was a brilliant scientist who discovered several very important laws of our universe. The story is that Newton, upon observing an apple fall from a tree, realized that the apple accelerated due to some invisible force. Newton's Second Law shows that the force of gravity acts on the apple to cause this acceleration.

Next, Newton brilliantly deduced that this force reached all the way to the orbit of the moon. He concluded that the orbit of the moon about the earth must be a consequence of the same gravitational force.

Newton's second law of motion - *The acceleration of any object is directly proportional to the net force applied to it and inversely proportional to its mass.* Or in other words, *force is equal to mass times acceleration.*

Newton's second law establishes a connection between an object's mass (m) and its acceleration (a), in the form of the equation $F = m \times a$ where F represents force, measured in newtons. What this means is that if you have a very small mass and you want to propel it at a very high acceleration, you will need a very strong force.

A good example is a bullet fired from a high-powered rifle. The bullet weighs much less than the gun, and so when the powerful powder explosion occurs, it is propelled at a very high velocity. Most of the energy of the explosion is changed into kinetic energy in the speeding bullet.

Newton's third law - *For every action there is an equal and opposite reaction.*

For every force applied to an object or surface, that object pushes back with equal force. This is what makes internal combustion engines like the one in your car operate to produce power. As explosions occur within the cylinder, they drive the piston away from the head, and this causes the crankshaft to turn and provide circular motion. The explosion pushes equally on the head and the piston, but since the head is stationary and relatively immovable, the piston is driven away with the whole force of the explosion. A small fraction of the force is stolen by friction.

This is also what enables missiles, bullets and space ships to fly. Combustible material is ignited in a partially enclosed chamber and this creates a violent but controlled explosion. This drives small particles of mass away from the explosion at a very high velocity.

In a gun or a rocket motor, the material can only exit in one direction. As it does so, the explosion produces an equal and opposite force in the other direction. This force pushes the gun back into the hands of the firer (recoil), and in the case of the rocket ship, it causes the space ship to lift off of the ground and accelerate as the burnt fuel particles are ejected downward. This principle is also used in space travel for retro-rockets, alignment rockets, and things like the Lunar Lander.

Even the electric motors in your gas-saving hybrid depend on this law: As magnetic forces are created in the fields; these forces pull the magnets in the rotor and create circular motion. And all cars, tanks and buses exert a force through friction onto the road surface that pushes them along.

It is also what propels airplanes of all kinds through the air. Planes with propellers and piston engines, as well as turboprop engines, are like huge fans that grab air from the front and accelerate it toward the back of the plane. Jet engines also take cold, fairly stationary air and accelerate it rearward at extreme velocities that push the plane forward.

On older jet engines, all of the air was passed through the combustion chamber. In many newer jet engines, part of the air is pushed around the combustion chamber by a ducted fan (propeller inside housing).

If these two natural laws (2nd and 3rd) did not exist or varied much from what they are, most of the machines and mechanisms on Earth would not work. There would be no internal-combustion engines, no trees swaying in the wind, no high-fives, no pitches or hits in baseball, nor many of the other things we see happen around us every day.

Section VII –

Chance or creation?

We are left with only two viable choices about our origins: Either we are a product of random chance as naturalistic evolutionists believe, or we were designed and made by some outside influence.

Either there is a cause or there isn't. We should not pass the buck like Richard Dawkins and say that aliens brought life to Earth. That still doesn't answer the question of whether or not there is a cause for the existence of our universe. In evolutionary theory, it is believed that the world and all life forms made themselves.

The belief in a Creator of some sort is based on the fact that nothing can exist without a cause and that nothing as complex as our world could come about through successive rolls of the dice.

The three elephants in the "science laboratory" are the questions: 1) Why is there something instead of nothing? 2) What is the origin of the energy from which our world is made? And 3) what is the reason for the extreme complexity of everything around us?

Questions about "how old is the universe?" or "how complex life works?" or "what's inside the atom?" are secondary at best. Analyzing how a clock works is fascinating, but knowing who made it and how it was made is the beginning of wisdom.

46.

The mathematics of probability

Statistics, likelihoods, and probabilities mean everything to men, nothing to God. Richelle E. Goodrich

Probability is the likelihood or chance that something will happen. It is a measure of how likely an event is to occur out of the number of possible outcomes. Calculating probabilities allows you to use logic and predict an outcome with some degree of certainty.

Probability is an estimate of the relative average frequency with which an event occurs in repeated independent trials. The relative frequency is always between 0% (the event never occurs) and 100% (the event always occurs). Probability gives us a tool to predict how often, on average, an event will occur, but does not allow us to predict when exactly an event will occur.

A simple example is the tossing of a coin. Since the coin is unbiased, the two outcomes ("head" and "tail") are equally probable; the probability of a "head" equals the probability of a "tail." Since no other outcome is possible, the probability is 1/2 (or 50%) of either a "head" or a "tail".

Thus, the probability of tossing a “head” is expressed as 0.5. An example of something that has a zero chance (0.0) of happening is that the sun will rise in the west. An example of 100% possibility (1.0) is that the sun will rise in the east.

If we ask someone to guess heads or tails for each of 100-coin flips, we'd expect about 50 of the guesses to be correct. Suppose a person walks into the lab and manages to guess heads or tails correctly for 60 out of 100 tosses. Is this evidence of precognition or perhaps the subject's possessing a telekinetic power that causes the coin to land with the guessed face up? No! In all likelihood, we've observed nothing more than good luck.

The probability of 60 correct guesses out of 100 is about 2.8%, which means that if we do a large number of experiments flipping 100 coins, about every 35 experiments, we can expect a score of 60 or better, purely due to chance.

But suppose this subject continues to guess about 60 right out of every hundred, and that after 1000 tosses, the subject has made 600 correct guesses. The probability of *that* happening purely by chance is less than one in seven billion, so it's time to start thinking about explanations other than luck.[8]

Ronald L. Wasserstein gives this example –

Here is an example I have used in classrooms all over the country, and it is way more fun than thinking about being struck by lightning! Imagine 175 million freshly minted one-dollar bills are being delivered to my house near Washington, D.C. One of those dollar bills is specially marked as the "lucky dollar bill." You get to pick a dollar bill, and if you happen to pick the lucky dollar bill, you win all the dollar bills.

A straightforward mathematical calculation using the dimensions of a dollar bill *reveals it will take two semi-trailers to deliver the 175,000,000 dollar bills to my house. How long will that line of dollar bills go if we lay them end to end?*

If we start from my house near D. C., we'll have enough dollar bills to go all the way south to Disney World in Orlando. Then we'll still have enough to go clear across the country to Disneyland! But, even then, we are not out of dollar bills, so we can go north and make it all the way to Portland, Oregon.

Still, we have dollar bills, enough to make it all the way east to Portland, Maine. And, fortunately, we'll have enough to make it back to my house near DC, completing the loop.

Do we have any dollar bills left? Yes! We would still have enough dollar bills to go all the way around the loop a second time!

Now imagine that you walk, bike or drive for as long as you want around the double loop, and when you decide to stop, you stoop over and pick up one dollar bill. Your chance of selecting the lucky dollar bill is one in 175 million, the same as your chance of winning the Powerball jackpot!

Naturalistic evolutionary theory says that cells came into being all by themselves without any outside intervention. However,

---

8 Some ideas taken from: guamet – Introduction to Probability and Statistics

the possibilities of this happening by chance are about the same as rolling a thirteen on a pair of dice.

The accidental formation of only one single protein out of the thousands of complex protein molecules that make up the cell is statistically impossible. Multiply this impossibility by itself as many times as there are proteins in a cell, and you will need to go to Staples office supply store to buy several more cases of little zeros to add to your numbers.

Proteins are giant molecules consisting of smaller units called amino acids that are arranged in many different but specific and precise sequences. The simplest protein is composed of 50 amino acids, but there are some that contain thousands of amino acids.

The crucial point is this: the absence, addition, or replacement of a single amino acid in the structure of a protein can cause the protein to become a useless molecular lump. Every amino acid has to be in the right place and in the right order. The theory of naturalistic evolution, that claims that life emerged as the result of random chance, is quite helpless in the face of this formidable impossibility since it is too wondrous to be explained by coincidence.

The fact that it is quite impossible for the functional structure of proteins to come about by chance can easily be observed even by simple probability calculations that anybody should be able to understand.

For instance, an average-sized protein molecule composed of 288 amino acids, and containing twelve different *types* of amino acids can be arranged in 10^{300} different ways. Of all of these possible sequences, only one forms the desired protein molecule. The rest of them are amino-acid chains that are either totally useless, or else potentially harmful to living things.

In other words, the chance probability of the formation of only one average-size functioning protein molecule is 1 in 10^{300}. The probability of this "1" actually occurring is nil. (In real life, probabilities smaller than 1 over 10^{50} are thought of as "zero probability").

How can scientists believe in this naturalistic evolutionary theory when it defies all the predictions of every bookie who ever lived? Why does naturalistic evolution require something that is considered impossible in the real world?

With the odds this low, this horse would never be permitted on the racetrack. It's worse than running a turtle against a cheetah in a hundred-yard dash, and then betting a 100 million Euros on the turtle.

47.

The statistical probability for life on Earth

Statistically, the probability of any one of us being here is so small that the mere fact of our existence should keep us all in a state of contented dazzlement."

Lewis Thomas

Life cannot have had a random beginning ... The trouble is that there are about 2000 enzymes, and the chance of obtaining them all in a random trial is only one part in $10^{40,000}$*, an outrageously small probability that could not be faced even if the whole universe consisted of organic soup.*

Fred Hoyle

Dr. Hugh Ross[9] has listed 68 necessary conditions for a life-sustaining planet to exist anywhere in the universe. By putting together the probabilities for each of these design features occurring by chance, he calculated the probability of the existence of a planet like Earth. This probability is very conservatively estimated at 1 chance in 10^{99} (it requires 99 zeros following the number - a million, for example, requires only six zeros).

Since there are estimated to be a maximum of 10^{23} planets in the universe (at 10 planets per star), by chance there shouldn't be any planets capable of supporting life in the universe (only one chance in 10^{76}, or 10 followed by 76 zeros). The national debt of the United States (in 2019) is around 20 trillion dollars – that is 20 followed by twenty zeros. Multiply that by 70 more zeros and we have the likelihood of the world forming by itself.

That is comparable to putting the names of the seven billion inhabitants of the earth in a big hat, and drawing out your own name 100 times in a row. It just isn't going to happen unless someone rigs the selection process.

Probability versus Statistics

---

9 *Big Bang Refined by Fire* by Dr. Hugh Ross, 1998. Reasons To Believe, Pasadena, CA.

Probability and statistics are related areas of mathematics that concern themselves with analyzing the relative frequency of events. There are fundamental differences in the way they see the world:

- *Probability* deals with predicting the likelihood of future events, while *statistics* involves the analysis of the frequency of past events.
- *Probability* is primarily a theoretical branch of mathematics that studies the consequences of mathematical definitions. *Statistics* is primarily an applied branch of mathematics, that tries to make sense of observations in the real world.

One of the adages of probability and statistics is: Any time something keeps happening that has one chance in twenty, or fewer, of happening, something is making it happen. In other words, it is no longer chance, but has an outside cause.

The casino owner's law says: *Anyone who wins a game more often than his chances of winning would reasonably allow, is cheating*. Over time, a player's actual winnings *must approach the calculated chances of winning*.

If they do not, the house should bar him from the gaming tables. *Winning streaks always break.* Casino owners should never believe in the Theory of Naturalistic Evolution because the odds are dramatically stacked against it.

What are the odds that our universe, our galaxy, our solar system, our sun and our planet all happened by chance and ended up the way they did? How many successive lucky throws would be required for our world to coalesce on its own?

How many more lucky throws would we need to make the first protein? How many more lucky throws are required to make a hummingbird? How many more would we need to make a human brain?

Our stainless-steel coin would have worn out long before we ever had a precise and benevolent universe like ours. We exist because someone rigged the process. Someone very smart and very powerful made it happen.

48.

How old is the Earth?

Age may wrinkle the face, but lack of enthusiasm wrinkles the soul. Ancient Proverb

In earlier chapters, we have looked at some of the scientific evidence for the age of the universe and the age of the earth. Science is sometimes based on assumptions that may or may not be true; that's why science is always in a process of change. Every scientist normally begins with basic knowledge discovered by other humans, and tries to build upon those prior discoveries.

Most scientists make a hypothesis and then try to find evidence that either disproves or supports it. "Science" is not an exact science, but rather a logical method that helps us discover answers to questions about our physical reality.

In the Physics or Chemistry lab, we were taught to take measurements at different times and in different situations and draw conclusions based on our findings. We were taught the process of "extrapolation." After taking measurements of some activity over the course of a small temperature range, we could then extrapolate those results to other temperature ranges.

Extrapolation means to "extend the line" that we have discovered into undiscovered or unproven territory. It is a commonly used and accepted practice when it is kept within reasonable limits. The danger occurs when we try to extend our limited data to a great distance beyond our empirical discoveries.

An example of when extrapolation does not work is in Newton's laws of physics. Einstein hypothesized that these laws begin to bend at extremely high velocities. So we have the Theory of Special Relativity that says that time stops at the speed of light, but that to get to the speed of light, we would need an infinite force to propel us to that speed, and that anything that was able to reach that speed would have infinite mass.

So, although Newton's laws are almost universally true, there are exceptions when we move into distant uncharted territory.

Several of our dating mechanisms for determining how old something is are based on extrapolations. We have taken readings over a relatively short period of time and then extended those findings back millions and billions of years into the past. It is the best method we have, but it is based on the unproven assumption that these processes have remained the same from the beginning of time.

The chicken or the egg –

If you believe in creation instead of random evolution, then you know that the answer to which of the two came first is the chicken. (In 2010, British scientists discovered that the chicken had to come before the egg, because the source of the protein that makes the egg shell is found in only one place – inside the chicken).

If a medical examiner could have examined the first chicken that was created, what age would he have put on it? If he came up with the right conclusions, he would have determined that the chicken was a year or more mature, but only a day old.

Someone has commented that it would have been deceptive of God to create the earth and the universe with age. However, if he had created Adam and Eve as babies, who would have taken care of them? Could they have survived?

If indeed there is a God who created humans and other animals and birds as adult specimens, then it was done out of necessity and not out of any intention to deceive us. If he created trees of sufficient age to produce seeds and carry on their progeny, why is that deceptive? If he created the earth with plate tectonics in working order, and a moon and a Jupiter all in place to fulfill the purposes for which they were made, how is that being intentionally deceptive?

If animals could have been created with adult age, could not the universe and everything in it have been created with age? What apparently is 14 billion years old may have been created much closer to our present time. Scientific evidence gives us some fair conclusions, but they are based on the assumption that all things began as an egg and not as a chicken.

For the earth to function as an ideal place for complex life to thrive, it had to have a thermostatic system to control its temperature. If you believe in a Creator, then you can understand that he would need to make the earth with all the plumbing systems complete and everything functioning *before putting life on it* – to do otherwise would have meant making an imperfect and dysfunctional world, and risking the survival of his creatures.

So, as believers in creation, we believe that God made the earth with a molten core, a magnetic field, a viable atmosphere, and with plate tectonics and volcanoes before he made the living creatures that depended on them – these mechanisms couldn't just appear gradually while life existed because they would not have been capable of maintaining a viable temperature and atmosphere.

If the Creator is as wise and powerful as we believe he has to be, then it would have been easy for him to create our world with age included. When he made a full-grown tree, it appeared to be 100 years old (along with its 100 tree rings), but it really wasn't.

What now takes 100 years to produce could have been completed in a moment. If it were to look like a 100-year-old tree and be able to reproduce itself, it would have had all of the adult systems in place. When he made an elephant, it was an adult elephant with the capability of reproducing.

If you believe that God created Adam and Eve as adult beings and not as toddlers, then it should not be so hard to believe that everything in this world was also created with a mature age.

The possibility of our world existing as a direct creation by a supernatural being changes many possible scientific conclusions, at least in the areas of time and the age of things.

If an all-wise and all-powerful Creator made our world, then the process or method that he used to do it is of secondary importance. Whether or not the moon was created out of nothing to be a Moon, or it was made by a perfect collision between two planets in our solar system, is not so important.

We can do our research and draw our conclusions based on what we find, but if there are possible factors that would turn our conclusions up-side-down, we should at least acknowledge them. We should not worship science as an inerrant god who makes no mistakes. If you do a search on Google about scientific theories that turned out to be wrong, you will be reading for a long time.

For instance, here are just a few:

- Vulcan was a planet that nineteenth century scientists believed to exist somewhere between Mercury and the sun.
- The earth was forever increasing in volume.
- The Martian canals were a network of gullies and ravines that 19th century scientists mistakenly believed to exist on the red planet. Scientists drew detailed maps tracing their paths, and soon wild speculation began on their possible origins and use.
- The aether, also known as the ether, was a mysterious substance that was long believed to be the means through which light was transmitted through the universe.
- Prior to scientists embracing the notion that the universe was created as the result of the Big Bang, it was commonly believed that the size of the universe was an unchanging constant—it had always been the size it was, and always would be.
- Spontaneous generation – was a belief regarding the spontaneous generation of complex life from inanimate matter, and taught that this process was a commonplace and everyday occurrence.
- The Hollow Earth theory that stated that the earth was composed of multiple layers and was hollow inside.
- Removing blood from a person will help make them well (This is probably what precipitated the death of George Washington).
- Stress theory of ulcers: As peptic ulcers became more common in the 20th century, doctors increasingly linked them to the stress of modern life. Medical advice during the latter half of the 20th century was, essentially, for patients to take antacids and modify their lifestyle. In the 1980s Australian clinical researcher Barry Marshal discovered that the bacterium *H. pylori* caused peptic ulcer disease.

We look back and laugh at the level of science that existed a couple of hundred years ago, but who is to say that in another two hundred years, scientists will not look back at us and marvel at how we could have believed in some of our "proven" axioms?

The point I am making is that science is in a constant state of learning and of self-correction. We might say that science is in a process of *evolution*, and that some of what we believe to be solid proof today will turn out to be a mistake tomorrow. We should not unequivocally commit all of "our marbles" to any one theory, because it is likely that the theory will one day have to be modified as new evidence is discovered.

Evidence tells us that the earth is much, much older than 10,000 years as some creationists claim, but if it were indeed created by the spoken word of God, than it could certainly be much less than 4.5 billion years.

If indeed there is a Creator, then all bets are off about the *exact* age of Earth and of our universe.

49.

Creation or spontaneous evolution?

Evolution is the greatest killer of belief that has ever happened on this planet. An evolutionist

Evolution

The word "evolution" means simply "change over time." However, the naturalistic Theory of Evolution of our universe and of life by sheer chance is an attempt to explain the existence of the universe without the necessity of a Creator. It attempts to answer the question "Why are we here?" from an atheistic perspective. Though it is only a theory, it has much too often been treated as an established law of science in the scientific community.

A Pew Research Center survey of scientists in 2009 found that 51% believed in some form of deity or higher being. 33% believed in God and 18% believed in a universal spirit or higher power. That figure is in spite of the fact that anyone who is not a firm "professor" of naturalistic evolution in most of our universities or science institutions is looked down upon as a fool, and finds it hard to get the promotions that atheistic evolutionists seem to achieve.

There is a strong social and professional pressure to align oneself with the prevalent school of thought in order to avoid being considered an ignoramus.

In another critical *survey (University of Chicago)* of 1,044 medical *doctors* in the USA, 76 percent said they *believe in God*, 59 percent said they *believe* in some sort of afterlife, and 55 percent said their religious *beliefs* influence how they practice medicine.

It is accepted by nearly all scientists that adaptations by living organisms to their environments can lead to changes that are called micro-evolution. Man can initiate or provoke these kinds of changes through selective breeding and can create different kinds of dogs, horses, chickens, etc.

However, naturalistic evolution claims that random chance will make new phyla (different basic body designs) when there is no evidence to support that claim. It claims that one-celled organisms eventually evolve by chance into giraffes, hippopotamuses and every other kind of animal.

The theory of random evolution ignores or rejects the scientific evidence of the last fifty years that clearly indicates a definite point of beginning for the universe. As shown earlier, the best scientific evidence points to the sudden inception of the universe at about 13.8 billion years ago, and the formation of Earth around 4.5 billion years ago.

This does not allow enough time for random evolution to produce any life at all through chance, let alone all of the species of living organisms that exist or have existed on Earth. Also, the overwhelming improbability of the existence of our "just right" universe and of advanced life without extremely intelligent outside intervention is so high that it stands way beyond the possible.

The evolutionary model also rejects older evidence like the Second law of Thermodynamics that declares that when left alone, without outside intervention, all things deteriorate to a state of greater randomness (i.e. chaos). Things do not automatically organize themselves.

Study shows that structures break down instead of spontaneously building themselves up into more complexity. As shown earlier, this is not just a theory, but is one of the most basic laws of science that has been proven in the laboratory and in real life innumerable times.

For random evolution to have been the cause of man's existence, living things would have had to evolve along three distinct parallel tracks:

1. The evolution of physical components into the building blocks of life – where a pool of lifeless, random elements came together and was transformed into molecules, then into living cells that could progressively transform into amino acids, then into nucleotides and then proteins. These physical proteins would have had to mutate spontaneously from a single cell to colonies of complex life forms and eventually to human beings.

2. The vast amounts of information contained in the design instructions of the first cell would have required the accidental evolution of a system of chemical "letters." These letters would also have had to organize themselves spontaneously into functionally useful "words" and "sentences" in order to produce the very precise written RNA and DNA instructions needed to make even the simplest one-celled organism. From there, the task would have only gotten more difficult for these one-celled organisms to randomly create more and more complex genetic codes needed for advanced life forms like human beings.

If we took a troop of 50 monkeys with computer keyboards and allowed them to type anything they wanted over a period of 50 years, we probably would not get one intelligible sentence, let alone the complete works of Shakespeare with zero typing errors.

A *genome* is the name given to an organism's complete library of deoxyribonucleic acid (DNA), the extremely precise chemical compound containing all of the instructions that determine what an organism is, how it looks, and how it functions. The human genome contains about 3 billion base pairs of information within the nucleus of *every one* of our body's cells.

We would expect that the DNA genome of a human being would be the most complex of all living things since humans are considered the most advanced and intelligent life form to have come into existence on our planet. However, that is far from the case. The Paris Japonica is a slow growing mountain flower that has a DNA that is 50 times as long as that of a human.

Dr. William J. Cromie, writing in the Harvard Gazette, has pointed out that a raspberry has only 8 percent as much genetic material as you or I. We are not surprised at that because raspberries aren't very smart or complex. However, an onion is also not very smart, but it has 12 times as much DNA as a rocket scientist.

The creature with the longest genome studied (until the octopus) is the *Daphnia pulex*, a microscopic, transparent water flea. It has about 31,000 genes — 25 percent more than we humans.

One of the most surprising features of the *D. pulex* genome is its compactness: in spite of being only 200 Mb in size (sixteen times smaller than the human genome that is 3200 Mb in size); its 12 chromosomes contain a minimum set of 30,907 predicted protein-coding genes, more than the 20,000–25,000 contained in its human counterpart.

It was recently found that an octopus has 33,638 protein-coding genes in its genome - 10,000 more than the genes found on the human genome.

A pond-dwelling creature called Oxytricha trifallax has 15,600 chromosomes: humans have 46.

The key to more advanced life forms is not in the quantity of information, but in the quality.

3. For random evolution to be a viable explanation for life, we would need to have progressed from a state of nothingness to a state of full consciousness and of personal awareness. One-celled amoebas do not have emotions or intellect, nor do they have a free will. So, we would need to have graduated from a complete unawareness of existence to a level of intellect, emotions and will that we see in human beings.

Descartes statement, "I think; therefore, I am," is true, but the reverse is not true: "I am; therefore, I think." Bacteria exist, but they do not think. Our solar system exists, but it does not think. For spontaneous evolution to be true, the ability to think would have had to evolve from an unthinking, impersonal state to what we have in humans who are reading this sentence.

There is no precise evolutionary explanation for individual personality in our genes, chromosomes or DNA. Two children of the same parents can have very distinct personalities even though their basic DNA and upbringing are nearly the same. There is something about each person that is not just biological, and not even based on genetic information, and that factor comes from some source unknown to science and even to religion.

If any one of these three parallel tracks failed to develop, our complex human life could not exist. This makes the spontaneous evolution from nothing to everything we see today at least three times as unlikely to have ever happened spontaneously.

The study of Earth's flora and fauna confirms that micro-evolution does exist. We only need to look at the quantity of dog types to see that very significant human-directed changes can and do give us very different looking canine types within the same species.

But micro-evolution is a very different process than the macro-evolution that some people accept by faith, because they do not understand the differences between them.

In both humanly-directed breeding and the process of natural selection, certain physical characteristics are advantageous for those individuals that possess them. Therefore, those characteristics are favored either by the environment or by potential mates (or by human breeders). The important thing to remember is that this micro-evolutionary process *reduces* genetic complexity by eliminating those genetic characteristics that are not selected.

The genome of a thoroughbred dog or horse is much less rich and varied than that of a mongrel or mustang. That is a major cause of the genetic diseases that purebreds often experience, such as relatively frequent hip dysplasia and the resulting osteoarthritis that German Shepherds and other dog breeds suffer. The English Bulldog, because of this continual inbreeding, is considered to be nearing the demise of its breed. A plain old mutt is still the healthiest dog around.

Macro-evolution, on the other hand, would imply a tremendous *expansion* of the gene pool from that of the original hypothetical ancestor to that of all living things. Macro-evolution is not micro-evolution expanded over time through tiny incremental changes. If it were, there would be so many millions of slightly different intermediate forms that it would be very difficult to discover in the fossil record where one species left off and the next one began.

Evolutionary scientists would not be constantly in search of "missing links" and celebrating every rare, possible find. The fossil record would not look like the "lawn of grass" that we actually find. Instead, it would take the form of a nicely branching "tree of life," with every twig tightly connected to increasingly larger branches that were eventually connected to the trunk. This is what evolutionary biologists tell us it would be if Darwin's Theory were correct. But, the fact is that whether we are talking about dinosaurs or possible human ancestors, we see a lawn rather than a tree.

Instead of the tree shape, in which different kinds of plants and animals gradually branch off from the main trunk of the common ancestor of them all, paleontologists find different organisms appearing suddenly in the fossil record, and then either surviving relatively unchanged or going extinct. They don't gradually and constantly change into something else.

This phenomenon is called stasis that means basically "staying the same." Darwin recognized this problem, but said that the many gaps would be filled as future digs and research uncovered the many missing links.

Instead of his prediction coming true, after the discovery of many millions of plant and animal remains over the past 150-plus years, scientists continue to seek those many missing links. When it became obvious that the fossil record was not going to change, some paleontologists, like Stephen Jay Gould, proposed a new model of evolution called Punctuated Equilibrium that was based on "saltational" (jumping or leaping) changes, which proposed that the changes were in fast motion over relatively short geological time periods.

The model's strength was that it actually matched the fossil record. Its great weakness is that it lacks a convincing mechanism or set of mechanisms that would make such large leaps from one kind of creature to another possible.

From a distance, without understanding very much about their complex differences in anatomy, it can sound logical to expect fish to evolve into amphibians that should evolve into reptiles that should evolve into birds and mammals.

However, the more you "look under the hood" and examine the many drastic differences in those organisms (such as the different number of chambers in their hearts, the many significant changes needed to shift from a cold-blooded to warm-blooded circulatory system, etc.) you will find it difficult to imagine a very convincing sequence of how those changes could have occurred through very lucky random mutations. If even some of those missing intermediate forms had been found, it would not be so out of the question to imagine this process. But?

Scientists cannot submit any convincing evidence that a set of ear bones evolved from the jawbone of a lizard; that a human evolved from a monkey, or a horse from a horsefly. No one has found a true missing link showing that macro evolution exists now or ever in the past.

Some supposed missing links have been fabricated by deception or creatively imagined to be something they were not and then hyped as proof of evolution until they were proven false or errant by careful scientific examination (Examples include the Piltdown Man, The Nebraska Man, Ernst Haeckel's embryo falsifications - which shamefully are still used in science textbooks today, and more recently Archaeoraptor)

Other "missing links" of one kind or another were not intentional falsifications, but rather examples of creative interpretation that were later proven to be poor science (Dubois' Java Man and Leakey's Lucy).

Why have so many hoaxes and careless mistakes trying to prove the Theory of Evolution been taken seriously by the scientific community and fed to the public as science? In 1859, in his book "*On the Origin of Species*," Charles Darwin said: *"Why, if species have descended from other species by insensibly fine gradations, do we not everywhere see innumerable transitional forms?"* (This is from chapter six entitled Difficulties on the Theory of Descent with Modification).

Scientists who believe in evolution have been searching for transitional life forms ever since the theory was propounded, but these "innumerable" missing links have not been found. Billions of dollars and many thousands of man-years of researchers' lives have come up with no indisputable sequence of intermediate forms.

This lack of concrete evidence has tempted researchers to fill the void and grasp at questionable evidence to support their beliefs. Fraudulent fossils have sometimes been fabricated and presented as transitional forms. Unfortunately, though these falsely claimed discoveries make the popular news, their recanting and disapproval by other researchers seldom gets any attention at all.

The Theory of Evolution was presented as a possible explanation for the existence of the many species and genera of life in our world. Darwin was a true scientist who did much field work and analysis, and proposed a hypothesis to explain his findings. If he were alive today, he would probably reject macro-evolution because the overwhelming scientific evidence today does not support it.

If only we had more open-minded and truly empirical scientists like him and fewer of the ideological and agenda-driven philosophers posing as scientists, our knowledge of nature would have advanced even more than it has.

Scientists tend to live up in the clouds and develop esoteric theoretical language, and have a hard time explaining clearly to layman what they are observing and concluding. Engineers, on the other hand, are tasked with bringing science down to Earth and applying science in practical ways. My attempt in this book has been to bring the truth of the heavens down to earth where most of us live.

Engineers know that things do not suddenly pop into existence for no reason. Whatever begins to exist has a cause.

The universe began to exist suddenly –

Until the middle of the last century, it was widely believed that the universe had always existed. However, since Hubble observed that the universe was expanding, and the discovery in the 1960s of cosmic background radiation, it has become the overwhelming consensus of modern science that the universe began to exist, and did so with a Big Bang (instantaneous explosion), some 13.8 billion years ago.

Evolution and the biblical account are not reconcilable –

Some people believe that God created the world through the process of evolution. However, macro-evolution is not consistent with belief in the Bible. Genesis says that God created the world in six days or eons (possibly time periods much longer than Earth days – the Hebrew language uses a single word *yom* for both of these meanings and thus admits both interpretations), and then he rested after he finished. This means that the creation of the world was complete.

Evolution, by its very definition, says that the creation of new kinds of creatures and macro-mutation is on-going. The Bible says the reverse: that we are now post-creation. The fossil record agrees with the Bible since no new body types or phyla have appeared in over 500 million years. The two beliefs are not reconcilable. You cannot believe the Bible and also believe in macro-evolution, either spontaneous or God-guided.

Intelligent Design and Naturalistic Evolution are not reconcilable –

You also cannot believe in intelligent design and also believe in random-chance evolution. Spontaneous evolution says that nature is all that exists, and that everything began from a single living cell that formed fortuitously out of some amino acids that happened to end up at the same place, under the right conditions and at the right time in a perfect universe. It also claims that while this was happening, the incredibly complex genetic code needed to sustain life was self-written.

Intelligent Design says that all living organisms were designed intentionally by a superior Being and did not just occur by chance. We are left with only two choices: We can believe in an incredible and statistically impossible series of miracles that happened by chance, or we can believe in a Supreme intelligence that designed it all on purpose. Which is more likely?

Evolution is not reconcilable with itself –

Since Darwin made his proposal for the origin of life in 1859, tremendous amounts of scientific discovery have come to light. Darwin saw similarities in the design of different birds and animals, and how some of them adapted to different climates and conditions. At that time his theory or hypothesis was a pretty good guess. However, in the light of all of the new information, it is now a very poor one.

As already shown, we now know that even single-celled microscopic organisms are extremely complicated, with thousands of moving parts and DNA strands whose massive amounts of information would overwhelm the capacity of many computers. If only one of the many key functions of a complex cell did not work, the cell would die.

If a gasoline engine has no spark, it doesn't matter if its fuel system is perfect. A car without a steering wheel or some other system of control would be a dangerous piece of junk on a highway and would only serve for pulling off parts to fix other vehicles. Everything has to work at the same time in exactly the right way, or the car does not function properly. Cars are far simpler than the simplest living cell and neither of them comes into existence by accident. Neither cars nor complex living cells evolve from nothing – they are designed and put together by intelligent beings.

Also, there is no way that this library of design instructions in the cell could have come together by random evolution. As mentioned earlier, DNA is a long thin strand of information within a cell that dictates how a cell is made.

There are about 3 billion base pairs of information in *each* human cell and those pairs fit into a space of just 6 microns across (there are 25,400 microns in one inch). If you stretched the DNA in one microscopic cell all the way out, it would be about 2 meters long and if all of the DNA in all of your cells were put together the strand would be longer than the diameter of our solar system.

If Darwin had known about the complexity and information-carrying capacity of DNA, he would never have proposed his theory of spontaneous evolution – he was too much of a practical scientist to have believed it. Unfortunately, Charles Darwin had no way of knowing of this complexity. He can be forgiven for his erroneous hypothesis based on very limited knowledge.

However, "scientists" today who have this more complete knowledge, have no excuse for believing in naturalistic evolution – it is a statistical impossibility, not even reasonable in your wildest dreams.

John Bohannon, writing in *Science* magazine states that a mere milligram of the DNA molecule could encode the complete text of every book in the Library of Congress and have plenty of room to spare.

In a new study, researchers stored an entire genetics textbook in less than a picogram of DNA—one trillionth of a gram. James Bannon, of Ark Invest, writes that DNA storage has a major advantage over hard disk drives: integrity. As a highly stable molecule, DNA does not break down over time. If kept in a cool place, it can store information for hundreds of thousands of years. And, it can store huge amounts of data in almost no space at all. One gram of DNA can store more than 2,000 terabytes (TB) of data (100 times the amount of data stored in the Library of Congress) in the space of a fingertip.

The capacity and complexity of DNA as a written code and as a storage device is beyond anything that we could have ever dreamed. It makes the Library of Congress look like a first-grade math table in comparison. If we were to rediscover the Library of congress a thousand years from now, would we theorize that it came together by chance?

I have a few copies of my books at the Library of Congress. I can assure you that those books did not just happen, but were the work of many hours of reflection, study and editing. My critics could say that I could leave out the word "intelligent," but they cannot deny that they came about due to intentional design.

The fact that naturalistic evolutionary theory became commonly accepted in the last hundred years led many scientists down a path of firm commitment to the naturalistic evolutionary paradigm. This makes it harder for them to revise their prior convictions even in the light of these newer discoveries that contradict the Theory. The scientists who do question the status quo often face discrimination and reduced opportunities for advancement.

Jonathan Wells, a scientist within the Intelligent Design movement, describes one Chinese biologist's questioning of Darwinism because of the discovery of how the Cambrian Explosion revealed the sudden appearance of almost all the major phyla and many major classes in the fossils of Chengjiang. These fossils strongly contradicted the gradualism that Darwin had proposed.

The Chinese scientist was surprised that none of the American scientists attending his lecture responded to his critique of macro-evolution. They asked him questions about specific fossils, but nothing about their impact on Darwinism. When Wells told him that they were perhaps just being polite, because most American scientists disapprove of criticizing Darwin, the Chinese biologist laughed and said: *"In China we can criticize Darwin, but not the government; in America, you can criticize the government, but not Darwin."*

Here are some of the most famous hoaxes or careless errors that Darwinists have tried to use to support the theory of spontaneous evolution:

Piltdown Man - (1912) was a notorious hoax perpetrated early in 20th century Great Britain, in which a medieval human skull was combined with the lower jaw of an orangutan and subsequently "found" in a gravel pit near the village of Piltdown, England.

Nebraska Man - (1917) was a false evolutionary model artistically created from a single pig's tooth. In 1922 Paleontologist Harold Cook found the tooth in Western Nebraska, USA in Pliocene deposits that were alleged to be 6 million years old. This "Nebraska Man" was one of the main reasons that evolution started to be taught in schools.

Before Nebraska Man, the theory of evolution had a hard time gaining acceptance in schools but such was the fame of the Nebraska Man that evolution became the accepted norm. Not long after it rose to fame, it was discovered to be a "dumb" mistake. The tooth belonged to a pig, not a man. This type of pig was found still alive in Paraguay in 1972.

Lucy (1974) – This fossil, discovered in Africa in 1974, was widely accepted by evolutionists and was the subject of some of the most intensive speculation about whether monkeys descended from us or us from them. Recently however, it has been revealed that Lucy (A. afarensis) had an anatomy ideally suited to climbing trees and was similar in many ways to other apes we are familiar with.

The French scientific journal Science Et Vie published the story in 1999 under the headline "Adieu, Lucy." One study, performed in 2000, discovered a locking system in Lucy's forearms enabling it to walk using the knuckles, in the same way as modern-day chimps. In a recent study, Tel Aviv University anthropologists determined that Lucy's lower jaw bone is similar to a gorilla's jaw bone. Other parts of the skeleton are just like the bones of knuckle-walking gorillas.

Haeckel's Embryo drawings (1866) – Despite the inaccuracies of Haeckel's drawings being exposed quite early in Europe, educators continued using the phony drawings for decades as proof of the theory of evolution. Astoundingly, they still appear as science in many US textbooks today (36 since 1985). Even worse than using the discredited drawings, they hide the fact that the whole recapitulation theory that they were based on has also been discredited among experts for many decades.

The continued use of those drawings in some text books to support Darwin's Theory is either a doubly false attempt to deceive the gullible, or an expression of extreme ignorance on the part of the educators who continue to publish those texts. The reason that this evidence is still being printed today is because it is the most convincing evidence they can find to support their religion of random naturalistic evolution.

Dr. Michael Richardson, an embryologist at St. George's Medical School in London, settled the matter of the drawing's authenticity once for all. He found there was no record that anyone ever actually checked Haeckel's claims by systematically comparing human and other fetuses during development. So, Richardson assembled a scientific team that did just that – photographing the growing embryos of 39 different species.

In a 1997 interview in The Times of London, Dr. Richardson stated: *"This is one of the worst cases of scientific fraud. It's shocking to find that somebody one thought was a great scientist was deliberately misleading. It makes me angry. ... What he did was to take a human embryo and copy it, pretending that the salamander and the pig and all the others looked the same at the same stage of development. They don't. ...**These are fakes.**"*

Archaeoraptor (circa 1990) – This was a forged fossil originally hailed as the missing link between dinosaurs and birds. A Chinese farmer who sought to make extra money by recovering and selling fossils created it. The farmer glued several fossils together, either out of the belief that they belonged together or a desire to make a more pleasing subject for display.

The composite fossil was purchased by a United States museum and then made its way to National Geographic in 1999. Despite various concerns being raised about the authenticity of the fossil and two peer-reviewed journals declining to publish a scientific paper about it, National Geographic proceeded with publishing an article claiming it was an evolutionary link between carnivorous dinosaurs and modern birds.

The Associated Press quickly picked up the story, and soon all the major news networks were reporting about this "fierce turkey-sized animal with sharp claws and teeth." Philip Currie of the Royal Tyrrell Museum in Alberta, Canada, and one of the scientists involved in the examination of *Archaeoraptor* for *National Geographic,* boasted: *"We're looking at the first dinosaur that was capable of flying."*

Eventually, the forged nature of the fossil was made public, and National Geographic was forced to "quietly" retract its claims.

We must look for consistency. Where there is a want of it, we must suspect deception.

Sherlock Holmes

If you search the internet for these famous fossils, you will find related articles still believing that these are great archeological finds that support evolution (I came across another one yesterday stating that Lucy was our human ancestor). You will find them in high school and college textbooks. You may find them cited on the Science Channel on TV. However, the professional paleontologist is now well aware that they are frauds. True science does not need to rely on fraudulent evidence to support its theoretical claims.

The missing link is still missing. In reality, whole chains are missing and not just one link. There is not an ounce of easily irrefutable scientific evidence to support the theory that life began spontaneously at any time in the past. There is no reasonable explanation for life's existence other than that someone made it.

Even Richard Dawkins, the famed proponent of godless evolution, once admitted, perhaps tongue in cheek, that life was probably brought to Earth by aliens from outer space. He refuses to believe in God, but admits that life is too complex to have made itself.

Other scientists have also proposed the same belief that life was seeded here from outer space, simple because they correctly find it impossible to believe that it could have happened by chance.

I believe that there are many scientists who are convinced deists like Albert Einstein, who believe in a brilliant creator, but not one who is personally related to his creation. Among many other statements to that effect, he said, *I believe in Spinoza's God who reveals himself in the orderly harmony of what exists, not in a God who concerns himself with the fates and actions of human beings.*

There are also practicing agnostics, who like Charles Darwin, believe that you cannot know God personally. Darwin made this statement: *The impossibility of conceiving that this grand and wondrous universe, with our conscious selves, arose through chance seems to me the chief argument for the existence of God; but whether this is an argument of real value, I have never been able to decide.*

I find it hard to believe that those who profess to be atheists are people whose refusal to believe in God is based on the scientific evidence. Some professing atheists have arrived at that belief system for personal, non-scientific reasons.

Madame Curie, the only scientist to win two Nobel Prizes in two different science categories, became an atheist or agnostic after the deaths of her mother and sister. It's possible that this loss influenced her theological beliefs.

Alan Turing, a mathematician and computer scientist, lost faith in God after the death of his first love. Turing could not understand how a real God would allow her to die. Turing therefore embraced atheism.

Several atheists I have known are angry (at God) because at some point in their lives, things did not turn out the way they wanted, and they believe it is God's fault. It may be due to the loss of a loved one, a strong desire they had that was never realized: or perhaps of a failed parental relationship. In some, perhaps it is just the desire to rule their own lives without God's interference.

Some atheists, at least on the internet, sound like very angry and bitter people. What is the origin of their anger toward God and toward those who defend belief in him? If they are sure that God doesn't exist, why get so bothered about a fairy tale?

Someone once asked the brilliant Albert Einstein what percentage of the world's knowledge he might know; *Was it 10%?* Einstein very humbly replied: *No, perhaps one tenth of 1%*. Today, with the modern age of internet information, it might seem possible for a person to know as much as two-tenths of one percent of all there is to know.

However, the amount of scientific information has grown exponentially since the time of Einstein, so the most knowledgeable person today probably knows an even lower percentage then Einstein did. When a person says that he knows that God does not exist, he is making a statement based on very limited information. How can he say that God does not exist if his own knowledge is so limited? Shouldn't he more correctly be an agnostic who says that if God does exist, he has not found him yet?

50.

Other reasons for believing in God

A God who let us prove his existence would be an idol."
Dietrich Bonhoeffer

As you have seen, there are many authentic scientific reasons to believe in God. No other explanation is reasonable other than that someone made this world and it did not come into existence by accident and then evolve all by itself.

There are also other reasons for believing in God that are not based on pure science but on observed statistics, human behavior, or the complexity of living organisms:

1. It has been observed many times over the centuries that *"Man is incurably religious"*

An opening statement of a recent CNN story read, "*Religion comes naturally, even instinctively, to human beings, a massive new study of cultures all around the world suggests.*" Here, we have the results of a serious study that says religion is not only here to stay, but is built into the very fabric of human existence. In 2007, the John Templeton Foundation made an award of £1.87 million to the University of Oxford for its Cognition, Religion, and Theology Project, and the conclusion was that religion and belief in God is an instinctive part of humanity.

This contradicts the secularization theory that held sway for nearly 200 years in Western intellectual circles. This theory posited that as societies modernize and become more scientific, people would become less religious.

Millions and millions of people from different eras and different cultures claim to have had an experience of the divine. The fact that all societies and cultures believe in gods of some sort, proves that man is naturally god-seeking. These religious tendencies were not propagated by missionaries or armies or via satellite.

Tribes that have never had previous contact with any of these things are all found to be religious to one degree or another. Their image of God or gods may seem very strange to us, but they all recognize that some deity exists.

2. Psychologists and sociologists assert that the person who believes in God is generally a much happier and well-balanced person.

This depends to a large extent on the image of God that a person has in his heart and mind. If a person's "god" is very strict, unloving, vengeful and cruel, then that person is not better off, but often lives in fear of what his "god" might do to him. However, if the person's image of God is one of love, benevolence, patience, and kindness, then that person is very much better off in his emotional well-being.

In Hebrews 11:6 we read, "*And it is impossible to please God without faith. Anyone who wants to come to him must believe that God exists and that he rewards those who sincerely seek him*" (NLT). It is of little help to believe in the wrong kind of "god." It is of very much benefit to believe in the God who is good, who rewards a person for seeking him, and who loves and wants the best for his creation.

If there is no God, there is no meaning, no purpose, no comfort and no source of lasting enjoyment. Nietzsche, who declared God to be dead, put his atheistic beliefs into practice in his life. Though he was a brilliant thinker and wrote prolifically for years, he descended into insanity, and died at the young age of 51.

The person who does not believe in God is a person with little hope for the future. If we are nothing but an accident of nature, then our life has no real purpose here in this world or beyond the grave. We would be not much better than a blade of grass that prospers for a brief time but then is soon forgotten.

Even if God did not exist, it appears that it is beneficial for man to believe in him. This is certainly not the reason that most people believe in God, but knowing that faith in God is beneficial should open people up to at least consider and investigate who he might be.

3. People who believe in God are healthier and live longer

Having enough faith to put your life into the hands of a higher power can be a freeing experience that allows you to relax and take what God gives you each day. The more you trust in the goodness and providence of a higher power, the less stress you experience and the less likely you are to suffer stress-related physical ailments that often bring on an earlier death. Christian values such as forgiveness and the feeling that life has an inherent purpose can also contribute to a person's lifespan.

An analysis of 42 studies involving 125,286 patients, published in the June 2000 issue of Health Psychology, found that those with some sort of religious involvement are 29 per cent more likely to live longer than non-churchgoers. A University of Texas study found that church attendees live up to seven years longer than non-attendees, and also that regular attendees have the longest life expectancy.

A similar health benefit was reported in the American Journal of Public Health, that studied nearly 2,000 older Californians for five years. Those who attended religious services were 36 per cent less likely to die during this half decade than those who didn't attend.

In 1990, the American Journal of Psychiatry discovered that believers with broken hips were less depressed, had shorter hospital stays and could even walk farther when they were discharged compared to their similarly broken-hipped and hospitalized, but unbelieving peers.

A Duke University study of 2391 people who were at least 65 years old found that regular churchgoers who also prayed daily or studied the Bible daily were 40 per cent less likely to have high blood pressure than those who did not. Elderly churchgoers had better mental health and were less likely to have high degrees of a protein associated with age-related illness.

4. The concept of God is necessary for restraining evil in human society.

Acknowledgement of God's existence and active involvement in his world is like a policeman in the conscience of mankind who says that a Higher Power will eventually judge us for whatever we do. If we believe that there is no God, then justice is whatever we chose to make it, and we can kill and steal and pillage to our heart's content, so long as we are stronger than our victims.

Without the knowledge of and belief in God, man might well have gone extinct many centuries ago because evil would have destroyed everyone. Understanding that there will be a day of reckoning when we face our Creator makes us all a little kinder, gentler and more tolerant of others.

The cruelty of atheism is hard to believe when man has no faith in the reward of good or the punishment of evil. There is no reason to be human. There is no restraint from the depths of evil which is in man.

The communist torturers often said, 'There are no God, no hereafter, and no punishment for evil. We can do what we wish.' I heard one torturer even say, 'I thank God, in whom I don't believe, that I have lived to this hour when I can express all the evil in my heart.' He expressed it in unbelievable brutality and torture inflicted on prisoners (Wurmbrand, *Tortured for Christ*, 34).

If God does not exist, objective moral values do not exist, and there can be no such thing as true evil, but rather only what some people choose to call evil. Since we all know that genuine evil does exist in our world, as evidenced by the cruelty of some human beings, that is evidence that good also exists.

If both good and evil do exist, then where did our recognition of them come from? God must exist as the reference point for determining which is which.

51.

Evidence for a Creator in living organisms

It is absolutely safe to say that if you meet somebody who claims not to believe in evolution, that person is ignorant, stupid or insane (or wicked, but I'd rather not consider that).
Richard Dawkins

The fact that life evolved out of nearly nothing, some 10 billion years after the universe evolved out of literally nothing, is a fact so staggering that I would be mad to attempt words to do it justice. Richard Dawkins

Well, what if I'm wrong, I mean — anybody could be wrong. Richard Dawkins

It could be that at some earlier time, somewhere in the universe, a civilization evolved by probably some kind of Darwinian means to a very, very high level of technology, and designed a form of life that they seeded onto, perhaps, this planet. Now that is a possibility, and an intriguing possibility.

And I suppose it's possible that you might find evidence for that if you look at the details of our chemistry, molecular biology, you might find a signature of some sort of designer, and that designer could well be a higher intelligence from elsewhere in the universe. But that higher intelligence would itself have had to have come about by some explicable, or ultimately explicable, process. It couldn't have just jumped into existence spontaneously. Richard Dawkins

It should be clear that this type of attitude is not science but biased, uneducated guesswork.

Up to this point, I have purposely avoided giving examples of living organisms to prove God's existence. The reason is that atheistic evolutionists claim that all life forms evolved through a series of billions upon billions of favorable mutations until they arrived at what they are today.

Since life forms are probably the greatest evidence for a Creator, I will make a brief list of some of the most outstanding creatures and show why they could not have evolved by random chance.

A. Irreducible complexity –

Dr. Michael Behe wrote a book in 1996[10], entitled *"Darwin's Black Box,"* in which he made the point that there are thousands of living organisms so irreducibly complex that there is no possible way that they could have evolved by chance. He used the example of a simple mousetrap to illustrate that if only one of the parts of a mouse trap is missing, the trap will catch no mice.

Behe's argument is that there exist thousands upon thousands of living organisms so complex that they could not have evolved by successive, gradual mutations. They simply would have lost the battle of survival since they could not function until they had managed to combine thousands of accidental mutations into a coherent functioning organism. Darwin made this famous statement: *If it could be demonstrated that any complex organ existed, which could not possibly have been formed by numerous, successive, slight modifications, my theory would absolutely break down. But I can find out no such case.*

One example of Darwin's "irreducible complexity" is the motor and propeller propulsion system, called a bacterial flagellum, found in many bacteria, including the common E. coli. The propulsion system of the bacteria has 40 moving parts made from protein molecules, including a motor, rotor, stator, drive shaft, bushings, universal joint, and flexible propeller.

The motor is powered by ions and can rotate at up to 100,000 rpm. It can reverse direction in only 1/4 of a revolution and has an automatic feedback control mechanism. The size is 1/100,000 of an inch (1/4,000 mm) in width, much too small to see with the human eye. This living motor is many times more complex than most if not all electric motors that man has ever made.

---

[10] *Darwin's Black Box: The Biochemical Challenge to Evolution* Michael J. Behe. 307 pp. The Free Press, 1996

Here are some other examples that Darwin never imagined of how complicated organisms function:

1) The ugly octopus –

In an article published by Discovery Institute (www.discovery. org), the authors point out that evolutionists are struggling to make sense of the "lowly" octopus. These creatures don't need prosthesis when they lose an arm, they just grow it back. They can change the color of their skin, or unleash a cloud of black ink, as well as fly through the water as if they were jet propelled. And it turns out that they are surprisingly smart.

Its genome was recently completed, and low and behold, an analysis of 12 different tissues from the critter revealed hundreds of octopus-specific genes found in no other animal, many of them highly active in structures such as the brain, skin and suckers. Also, the octopus genome turned out to be almost as large as a human's and to contain a greater number of protein-coding genes -- some 33,000, compared with fewer than 25,000 in Homo sapiens.

In past decades, the common explanation for mutations and new functions for old organs was that genes were duplicated and then the copy went on to mutate and serve different purposes.

Evolutionists believe that the selective process can only act on the genetic material that is available to it, and that it relies entirely on random mutations of older cells to provide new building materials. Now, however, the octopus sticks a tentacle in the gears and shows that it has a lot of genes never found anywhere else.

The solution that evolutionists come up with to explain this deviation is that these new genes are "de novo" (out of nowhere). The problem is that *de novo* isn't any explanation at all, but acts like a magic wand to be used when evolutionists encounter a unique gene and have no way of explaining its origin via the method of gene duplication that was their stock in trade until recently.

It seems like we are returning to the science of alchemy in which we can magically turn lead into gold, or even make gold out of nothing. Whenever you see the term *de novo* invoked, you can be sure that the evolutionary biologists lack any scientific explanation for that gene's origin.

2) **The Bombardier Beetle –**

This beetle is a marvel of nature, a sort of six-legged tiny tank. It defends itself by mixing chemicals that explode when combined, then firing the boiling explosive through twin tail tubes that can swivel like gun turrets. The chemicals produced inside the beetle react and shoot out at temperatures that burn the beetle's enemies. The force of the shot fired should be enough to blast the little beetle yards away, if not into pieces.

The secret to its un-cannon-y ability to shoot and not recoil is that the "crafty" beetle actually lets go with a stream of up to 1,000 miniature multi-explosions. Our armed forces pride themselves on having the mini-gun, but this beetle had it long ago.

3) **The Giraffe –**

The giraffe's heart creates immense pressure to drive blood up the neck to its brain. Because of this there are valves in the neck that automatically restrict the blood flow when the giraffe lowers its head to drink (as much as 22ft difference in height).

Without these valves the sudden increase in blood pressure as the giraffe lowers its head would rupture the arteries in the brain and kill the giraffe. However, the giraffe could not have evolved a long neck without the valves, and would have had no need to evolve the valves unless it had a long neck.

4) **Megapode or "incubator" bird of Australia –**

This very unusual type of turkey-like bird weighs three to four pounds and is unique among birds. All birds, except the incubator bird, use body heat to incubate their eggs. Instead, this innovative creature piles up great piles of organic debris that serve as the incubator. The warmth produced by the rotting compost provides the heat needed to hatch the eggs.

Once the nest passes the test, the female lays 20 to 35 eggs, one every three days for up to seven months. Each egg weighs about a half a pound and is as large as an ostrich egg – laying each egg is a formidable feat for a three to four-pound hen. Once the eggs are laid, the hen leaves the nest and never comes back to help in the incubation process or the raising of her chicks.

Now, it is the male's turn to do some work. His role is to manage the temperature and humidity level of the incubator mound in which the eggs are deeply buried. For the eggs to hatch, they must be maintained at an "eggsact" temperature of 91°F. This temperature may not vary more than one degree to either side of 91° F. It is surmised that the male bird has a thermometer in its beak: other researchers think it may be in his tongue.

The male digs down into the nest and checks the temperature. If it gets too hot, he may pile lots more debris over the mound to protect it from the sun. Often, he even rearranges the whole pile of rotting leaves and other organic material.

Not only must he maintain the temperature, but he must also keep the humidity of the pile at 99.9%, or the eggs will dehydrate and kill the chicks inside. If this bird's thermometer were off by a degree or more, its species would have gone extinct long ago. Everything has to be just right all of the time or it could not survive.

5) The hummingbird

The tiny hummingbird weighs less than a nickel. Its rate of metabolism is so high that during the day it needs to continually be eating/drinking to keep from dying. It has no rods in its eyes, so its eyes cannot see in low levels of light. At night, it goes into a state of hibernation. But this pint-sized critter is 100 times technically superior to any flying machine that man has made. Consider these facts:

- The quill in the feather is stronger for its weight than anything designed by man.
- The quill constantly changes shape to meet the requirements of the air pressure and wing position.

- Its flexibility allows the feathers at the tip to bend upward with each downbeat of the wing. This is equivalent to changing the pitch of a helicopter blade.
- The bird has enlarged muscles to operate the wings – they compose nearly ¾ths of its total weight.
- It has a remarkable system of respiration in which the hollow bones provide an air sac that provides buoyancy; a reservoir for breathing and an air conditioner.
- Air flows into the lungs in only one direction. This provides a continuous supply of necessary oxygen.[11]

B. The human ear bones –

The human ear contains three tiny bones that transmit sound waves from the eardrum to the cochlea (see more information in Chapter 42). Because of the complicated arrangement of those bones, transitional forms (that have never been found) would have served no purpose. If the relationship between them is off by a small percentage, there is no hearing.

Evolutionists claim that these bones used to form part of the reptilian jaw joint, but no intermediate fossil with half jaw/half ear (or any combination thereof) has ever been found, and there is no reasonable way evolution can explain how a jaw becomes a sensitive, functioning ear bone. The closest anyone has come to explaining it is to say, like my deceased father-in-law, that someone had talked so much they had *"jawed his ear off."*

C. Metamorphosis –

There is no credible explanation within the theory of evolution for the mechanism by which a caterpillar turns into a butterfly. Essentially the original animal is broken down completely and rearranged into a totally different animal, with no similar

[11] Some ideas taken from *The Revolution Against Evolution* by Douglas B. Sharp

physical characteristics. This process cannot be explained within the theory, and must have some other explanation.

Evolution is primarily about gradual changes in physical life forms and behavioral habits, but it cannot explain something as weird as a complete realignment of the cells so as to form a wholly different creature.

Some scientists have posited the idea that the caterpillar uses one part of the creature's DNA, and the butterfly uses another part. But they cannot explain why the caterpillar feels the need to enter the cocoon, or why it is always a butterfly that comes out of it instead of some grotesque mutation formed from random bits and pieces of its DNA. The whole process is too precisely controlled to believe that it just started to happen and keeps on happening by chance.

D. Close only counts in horseshoes –

One obvious fallacy in the Theory of Evolution is the belief that organisms adapted organs, appendages, bones, muscles, ears and eyes through a series of mutations over millions of generations. Each generation kept mutating just a little more in the right direction until finally the desired result was achieved.

Once it was achieved, the mutations apparently stopped since no new body designs have appeared in over 500 million years. Snakes still crawl, fish still swim, birds still fly and men still think in the same way as before.

Evolutionists theorize that rodents learned to fly by jumping out of trees, that insects grew four wings that are extremely complicated mechanisms and then kept trying to fly until they got the hang of it. They believe that around 400 million years ago, some fish began to walk on land, using their fins as legs. These land explorers then gave rise to all of the 4-limbed land animals, from lizards and frogs to birds and bears. Reptiles began to grow complicated feathers even though they didn't need them for anything, and indeed they would have served no useful purpose for millions of generations. If feathers were so wonderful, everybody, not just birds, would have some.

They also conjecture that reptiles began to convert legs into wings. However, legs that became lighter and weaker wouldn't provide any advantage to the reptile. Instead, it would be a disadvantage, and faster and stronger creatures would eat them. A half bird creature with half a wing would still be too heavy to fly and too slow to escape its predators.

The fact is that if a gear is missing from a watch, it cannot keep time. If a link is missing from a chain, it won't serve to pull anything. If a stomach is missing from a dog, it will die. If the human heart is missing a valve, *adios amigo*. It doesn't help to be *almost* good enough to accomplish something, especially when your life depends on it. The benefit of being *almost* able to fly for millions of years is a no-brainer.

If a wing is missing from an airplane or is not perfectly balanced from left to right, it will not fly. I was a pilot for several years, flying in the USAF in Vietnam and other places, as well as in single engine planes in the mountains and lowlands of Bolivia in the days long before GPS.

I can assure you that the wings of my plane had to be just right, or I would have crashed. If one wing is bigger than the other, or too short, or the wrong shape, the plane never gets off of the ground, or it crashes immediately after take-off. Everything has to be in perfect balance or you're not coming back alive.

E. The theory of junk DNA is no more –

For several decades, evolutionists explained away the complexity of DNA by saying that a large percentage of it was junk DNA, meaning it was just trash that had accumulated due to the evolutionary experimentation process and that it was of no use. *These regions have traditionally been regarded as useless accumulations of material from millions of years of evolution ... In humans, about 97 percent of the genome is junk.*[12]

With the termination in 2007 of the ENCODE (Encyclopedia of DNA Elements) project that mapped out the entire human genome, it was found that at least 80% of human DNA serves a useful function, and the old idea of DNA being a virtual junkyard had to be thrown on the scrap heap.

---

[12] Philip Yam, "*Talking Trash*," Scientific American, 272:24 March, 1995.

The conclusions of the project were so devastating to the evolutionist camp that many of their comrades rose up in opposition to accepting it. That opposition continues today. One of the rank and file in opposition to ENCODE's conclusions went so far as to say, *"If ENCODE is right, then evolution is wrong."*[13]

Numerous evolutionists who believe that their previous conclusions can stand only if ENCODE is proven false have careers, reputations, and deeply held worldviews founded on the premise that humans were created by random processes that would have filled our genomes with useless bits of DNA. Unfortunately, human nature normally doesn't allow people with so much face to lose, to concede defeat.

F. Instinct –

Evolution cannot explain how animals got instinct. Evolutionists say that animals and birds have learned to adapt to their surroundings and have found ways to survive. However, this cannot explain the fact that many birds fly thousands of miles to get to a place they have never been before.

If humans are the highest level of evolution, why do we still get lost in the parking lot while these birds can unerringly fly thousands of miles to a place they have never been?

Look at these interesting facts about bird migrations:

Nearly 40 % of the bird species living in Europe and Asia are migratory birds. This means that 3 to 4 billion birds are leaving Eurasia in the fall for warmer habitats. Each bird species has its own destination, its own route, with its own departure and arrival dates, its own way to travel (making many stops or flying as far and quickly as possible with nearly no stops) and its own habits (flying alone or with a group).

---

[13] *Dan Graur, "How To Assemble a Human Genome?" (2013).*

• Bar-headed Geese fly over the Himalayas twice a year and go as high as 7000 meters (23,000 ft) to get over the mountains.

• The Arctic Tern, a bird weighing about 100 grams, each year makes the round trip between the Arctic and the Antarctic (about 20,000 km or 12,000 mi) in order to never experience winter! They were doing this long before Europeans ventured out to discover America.

• The Sedge Warbler can fly 2400 miles without any stopover. This bird weighs 23 grams when it begins the journey and only 9 grams when it arrives!

• Other small migratory birds can fly for 75 hours straight at 3,000 ft altitude to cross the Mediterranean Sea or the Sahara Desert.

• One of the tiniest migrants, the Ruby-throated Hummingbird, flies 900 kilometers (550 miles) nonstop across the Gulf of Mexico using the benefit of flying in formations.

• The Bar-tailed Godwit was named the "endurance champion of the animal kingdom" because it completed a nine-day nonstop 6600 mile flight across the Pacific Ocean. Once the chicks are hatched and feeding on their own, mom and dad fly back to the southern hemisphere for the winter.

When the chicks have developed their wings and feathers, they also make the journey south. The amazing thing is that they have never made this trip before. And they do not have an experienced bird to show them the way. Somehow, they just know where to go.

• Pacific Golden-Plover – These birds make the 3,000-mile trip from Alaska to Hawaii in 50 hours. They breed on the Alaska tundra, and then abandon their chicks as they head for the golf courses and lawns of Hawaii. In this case, it is the chicks who suffer from the empty nest syndrome. They leave about a month after their parents and head south.

The birds speed along, between 50-60 mph, on the nearly 2,500 miles from their breeding grounds in Alaska to the Hawaiian Islands in about two days, nonstop. They flap their wings more than half a million times during the trip and carry just enough fuel in the form of body fat to reach their destination. They lack the waterproofing on their feathers that seabirds have, so they can't rest on the ocean. Recently, one bird was tracked by satellite as it flew from Alaska to Okinawa, a distance of 5468 miles (8800 kms) in 8 days.

- The families of most species of hummingbirds raise their offspring until they are out of their nests and able to feed themselves. A few weeks later, the parents disappear. The chicks are left alone to fatten up for their long migratory journey to a place in the tropics where they have never been. How do they know when to leave, how to get there, and when they have arrived?[14]

Many species of fish also migrate each year –

California Grunion – One very interesting fish knows exactly when to lay its eggs so that the tides will not wash them away. There is only one perfect night each month when these fish place their eggs on the sandy beaches of California in order to have enough time for hatching. Grunion mating occurs three to four days after the full or new moon, just after the highest high tide.

Once ashore, females wriggle themselves into the sand, tail first, and deposit up to 3,000 eggs. One or more males then wrap themselves around the female and emit sperm that flow down to the eggs and fertilize them. The female then frees herself from the sand and, along with the males, returns to the sea.

While the eggs may remain near the surface, the sand from the receding tide may also bury them up to 20 centimeters deep. Despite being on land, the developing embryos stay moist in the sand. About ten days later, rising tides cause the eggs to hatch, free the hatchlings and carry them out to sea.

---

[14] Some information adapted from:http://www.educapoles.org/news/news_detail/how_do_migratory_birds_find_their_way

G. For evolutionists, sex is a dirty word -

Sex is the queen of problems for evolutionary biology. Ninety-nine percent of multi-celled animals use sex to reproduce. They form eggs and sperm (gametes), mix those together and create offspring with an entirely new genome. Scientists have long wondered what processes caused sex to evolve and become so incredibly prevalent in the animal world.

The need to mate male and female pairs of a species together makes procreation much more difficult than when it can be accomplished asexually or through simple cloning of itself without the need of a partner. The "evolution" of sex has long puzzled biologists since its disadvantages seem to far outweigh its benefits.

Asexual reproduction passes on 100% of an organism's genes, whereas sexual reproduction only passes on 50%, which is a tremendous loss to the reproductive power of a species. Add to this the great effort expended in the need to find a mate.

Evolutionary biology is unable to justify how natural selection would make this giant step "backward" wherein reproduction is probably four times more difficult. It cannot explain how eggs and sperm conveniently came to each share exactly one half of the genetic material needed for continuance of the race.

Sex takes much longer and requires much more energy than simple cell division. Why would a process so unprofitable to a species become so widespread?

The usual evolutionary answer given for the existence of sex is that it helps eliminate damaging mutations accumulated through asexual reproduction. However, the disadvantages for a species that "wants" to continue to exist far outweigh the advantages.

Sir John Maddox, the long-serving, pro-evolutionist and distinguished editor of *Nature*, has stated, *"The overriding question is when (and then how) sexual reproduction itself evolved. Despite decades of speculation, we do not know. The difficulty is that sexual reproduction creates complexity of the genome and the need for a separate mechanism for producing gametes.*

The metabolic cost of maintaining this system is huge, as is that of providing the organs specialized for sexual reproduction (the uterus of mammalian females, for example). What are the offsetting benefits? The advantages of sexual reproduction are not obvious."[15]

Dr Carl Zimmer adds this: *Sex is not only unnecessary, but it ought to be a recipe for evolutionary disaster. For one thing, it is an inefficient way to reproduce.... And sex carries other costs as well.... By all rights, any group of animals that evolves sexual reproduction should be promptly outcompeted by nonsexual ones. And yet sex reigns... Why is sex a success, despite all its disadvantages?*[16]

If two sexes are much better, why not evolve into four or six or eight? This would do an even better job of eliminating unwanted mutations, but don't hold your breath waiting for random selection to produce it. There are a very few creatures that appear to be male and female as well as asexual, but apparently that system has not caught on.

H. The annual rate of extinction of species tells us that species are disappearing instead of appearing. Since the Cambrian Explosion (543 million years ago), there is no record of any new phyla (body design) appearing in the fossil record, but many known species from that period have gone extinct.

Since some phyla are disappearing and no new ones are appearing, extending the amount of time for evolution to produce more and more complicated organisms would mean that instead of a net increase in phyla, the longer time would lead to a net decrease. It would mean that less body designs are becoming available as time passes.

The "Cambrian Explosion" refers to the appearance in the fossil record of all or nearly all major animal body plans about 543 million years ago. The extremely large number and range of new fossil types appear in an interval of about 10 million years or less. On an evolutionary time scale, 10 million years is a "blink of the eye" that appears to be in contradiction to the expected gradual development of life forms.

---

[15] *What remains to be discovered,* ISBN 0-684-82292-X
[16] *Evolution: The Triumph of an Idea, 2001*

This very rapid rate of the appearance of fossils was already causing much debate as early as 1840, and in 1859, Charles Darwin suggested that it was one of the main objections that could be raised against his own theory of evolution by natural selection. If naturalistic evolution were indeed happening, then the rate of appearance of new phyla should be increasing, whereas in fact, it has stopped.

I. Homosexuality –

If homosexuality were a result of evolutionary genetics as some scientists claim, evolution would have led to the extinction of this gene. Homosexuals cannot reproduce offspring and therefore cannot propagate their "race." If homosexuality were an evolved trait, it would have led to its auto-extinction because a species that cannot reproduce itself will lose the battle for survival to the species that can.

J. Two kidneys and two lungs –

The existence of two symmetrical kidneys and lungs in humans and many other animals is unnecessary. If evolution only allows the survival of parts that are functional, then the existence of duplicate kidneys lacks a plausible evolutionary explanation. This is not the same for two eyes, or two ears because both of those organs work in coordination with each other and each serves a useful purpose that would not be achieved if there were only one. An animal with only one functioning kidney or lung has all the capacity he needs.

K. Pleiotropy –

Pleiotropy means that when one gene in an organism is changed, it usually has multiple effects on the organism. This renders the "improvement" of animals by random mutations very unlikely since any mutation with a potentially beneficial effect will also produce one or more potentially harmful effects.

What this means is that when a gene is varied in an organism, it usually causes more harm than good. When we take a wheel off of a car, it doesn't drive better. A one-winged bird doesn't fly at all. A fish with legs doesn't swim better than the larger fish that is trying to eat it. If macro-evolution were to tamper with the basic pattern of an organism, it probably wouldn't function better, it would die.

A fundamental principle in evolution is that in order to produce a new system, evolution needs to try lots of new things. It must produce multitudes of variations upon which natural selection can act in order to "find" something useful to retain. But in the scenario proposed by Darwinian Theory, you'd end up with an organism loaded with sub-optimal or useless parts.

Many evolutionary biologists say it's easy to evolve new structures, but as a result, lots of junk accumulates. The authors of one paper observe that this sort of reasoning *"has historically led evolutionary theorists to expect, that living systems carry a significant fraction of vestigial, or nonfunctional, elements, as well as quasivestigial elements which function with much less than optimal efficiency."*[17] This was probably the rationale for coming up with the idea of junk DNA.

There is a high energy cost involved in creating greater complexity. The cost of carrying along useless or redundant systems is one of the arguments used to explain why organisms would need to shed excess baggage. To test this evolutionary expectation, Snoke, Cox and Petcher constructed a model that rewards evolving a new function, but exacts a price for evolving systems that require lots of parts before providing an advantage.

Using this model, they found that when the model was setup to reward the evolution of new features, it did indeed evolve new features. Some of those features were useful, but the vast majority was not. It then happened that before the simulation finished, the population experienced a crash. The organisms accumulated so much genetic garbage -- new features that were in fact no more than useless freeloaders -- that their fitness dropped precipitously. In other words, along with good mutation comes

[17] David W. Snoke, Jeffrey Cox, and Donald Petcher, "Suboptimality and Complexity in Evolution," *Complexity*, (July 1, 2014).)

greater bad mutation and that eventually does more harm than good.

L. The complexity of coding in the cells –

In Darwin's day, science had not really progressed very far compared to where we are today. Darwin observed only outside physical characteristics of birds, animals and fish, but he could not see into the inner workings of their nervous systems, their circulatory systems, nor their immune systems.

Today we can look at many more of these things at the molecular level and see how remarkably complex they are. Darwin never dreamed of anything remotely like DNA.

As I already mentioned, DNA is a very fine, spirally coiled, double helix thread found in every cell of every living organism. It serves as an extensive guidebook or instruction manual on how to make different living cells. A complete blueprint of every individual human being is present in every one of its thousand-million-million cells.

What is amazing is that within a tiny space within every cell in your body, this code is *three billion letters long*. The human genome, or DNA strand, contains 23 DNA molecules each containing 500,000 to 2.5 million nucleotide pairs. You have about 75 trillion cells in your body, so if you stretched the DNA from all of the cells out end to end, they would stretch 10 billion miles, more than the distance from here to Pluto and back.

MIT-trained scientist Dr. Gerald Schroeder states - *Other than sex and blood cells, every cell in your body is making approximately two thousand proteins every second. A protein is a combination of three hundred to over a thousand amino acids. An adult human body is made of approximately seventy-five trillion cells.*

Every second of every minute of every day, your body and everyone's body is organizing on the order of 150 thousand, thousand, thousand, thousand, thousand thousand amino acids into carefully constructed chains of proteins. Every second; every minute; every day. The fabric from which we and all life is built is being continually rewoven at a most astoundingly rapid rate.

The DNA replication includes a built-in error checking method and a DNA repair process that works to prevent mutations. The fact is, any slip-up by the DNA leading it to mutate is almost always stopped or reversed.

To date, no cellular life has been found without DNA. Cellular life includes all bacteria, archaea, and of course eukaryotes that include multi-cellular organisms like us.

Perry Marshall, an information specialist, comments on the implications of this. *"There has never existed a computer program that wasn't designed; whether it is a code, or a program, or a message given through a language; there is always an intelligent mind behind it."*

Try to tell Bill Gates, the co-founder of Microsoft, that he no longer holds the patents for Windows or any of their other programs. These programs apparently just evolved with no input on his part, and therefore he has no right to charge a fee for them. Fortunately for us humans, God doesn't charge a license fee to use his design codes.

M. Seeds – precursor to Xerox –

In the late winter, we may go to a hardware store or department store and in the section of garden supplies we will find a display with packets of a great variety of seeds. Each packet usually has a picture of what we should get if we plant the contained seeds in the right way. We can choose flowers or vegetables.

How do the packagers know how the seeds are going to turn out; because seeds reproduce themselves as they are. When you sow a seed, you get a copy of its genetics. Every seed reproduces itself. We don't get peas from carrots, or rutabagas from sunflowers.

Every cell of every seed contains all of the information necessary to recreate itself. Though geneticists can successfully create hybrid seeds, the seeds produced by those seeds are usually less fertile and often produce a mixture of inferior varieties.

A Xerox copy machine whose design was produced at great effort and expense is able to make copies of documents, but it cannot make copies of itself that have the ability to make more copies of themselves ad infinitum.

In the imaginary world of Star Trek, they have "replicators" that supposedly produce things out of nothing, not even needing raw materials. But in the real world, we have tiny seeds that reproduce plants and seeds based on their own genetic codes, and with the ability to reproduce themselves.

Who are the true scientists? –

Atheistic evolutionists accuse scientists who believe in God or in Intelligent Design of being religion promoters and not true scientists. A fake scientist is one who enters into his investigation with a preconceived bias against finding a certain result. In spite of evidence to the contrary, he rejects anything that does not fit into his belief system. Thus, he ceases to be an open-minded investigative scientist in search of truth, and becomes a person trying only to defend his prior established positions.

The evolutionist's concept of God is one of a "god" so puny that he wouldn't be capable or interested in creating the vast variety of life forms that exist on our planet. They believe that since life is so amazingly diverse, it must have gotten that way by runaway mutations and not by intention. It is no wonder they refuse to believe in God if their concept of him is that small.

There are theoretical scientists and practical scientists. In the theoretical sphere, just like in Hollywood, everything is possible; but in the practical arena, many things are impossible. I am by training an engineer, which places me in the field of practical science. We deal in the real world, not in an illusory one that does not restrict itself to known physical laws.

Sometimes theoretical scientists and mathematicians are able to theorize in ways that expand our scientific horizons and our known physical laws, as Albert Einstein did. But after he proposed that $E=mc^2$, other more empirically-oriented scientists had to seek physical proof of his equation.

We need both kinds of scientists, but we should also remain keenly aware of the difference between that which is only theorized and that which has been empirically demonstrated.

As we have seen, the evidence for design or input from a master architect in the fabric of our universe is overwhelming. We are not here by chance. The world, and everything in it, has a purpose. To adamantly deny this "possibility" negates the privilege of calling oneself a scientist. All of the people in prison today have been convicted with less than 1/100th of the evidence for God's existence.

If it is true that we are biased on the side of believing in a Creator, staunch atheists are a dozen times more biased in categorically denying the possibility that God exists. To declare oneself an agnostic (one who believes that God might exist but doesn't know him yet) after seeing much of the evidence is acceptable: to declare oneself an atheist, saying that no God has ever existed nor can exist, is neither logical nor scientific.

I would challenge those "scientists" to open their minds and start looking for and accepting whatever truth they find and to be willing to reform their beliefs in the light of the evidence, and not vice-versa.

The best evidence that science has been able to come up with shows that our single universe had a beginning and that when it began, there were inviolable laws and rules already set in place that made complex life able to function.

An infinite number of rolls of the dice does not make the impossible possible. Imagining that there are an infinite number of universes has nothing to do with science but everything to do with science fiction.

H. G Wells, George Lucas and Gene Roddenberry have created imaginary worlds and unrealistic expectations of what life will be like in the future, but if you look for their books and films in the library, you will still find them in the section under "Fiction." Don't let imagination be sold to you as a substitute for science.

Section VIII –

Why consider God?

Chuan Tzu and Hui Tzu were standing on the bridge across the Hao River. Chuang Tzu said, "Look how the minnows are shooting to and fro! How joyful they are!"

"You are not a fish," said Hui Tzu. "How can you know that the fishes are joyful?"

"You are not I," answered Chuang Tzu, "How can you know I do not know about the joy of fishes? . . . I know it from my own joy of the water"

An ancient Chinese story

Atheists often affirm with great conviction that God does not exist, although they cannot point to any specific scientific evidence to support their claims. Some of the common reasons they give are:

1. This world is full of evil and therefore God is either not good or not mighty. They believe that if God cared about us, he would immediately take away all suffering and pain.
2. There is no evidence for God's existence. What they mean is that they have never seen God and therefore he must not exist. But dare they make the same assertion about the stuff inside their neighbor's attic?
3. God has never contacted me; therefore, he must not exist. That sounds like they want God to bow to them while they refuse to bow to him.
4. If God created the universe, who created God? They mistakenly believe that the maker of a watch was a clock and not a watchmaker. They expect the Creator to be of the same substance and on the same level as his creation. All things within our universe have causes, but God is both immanent (within our universe) and transcendent (above and beyond our universe.) He does not require a prior cause.
5. There is as much evidence for believing in God as there is for believing in the Tooth Fairy, or Santa

> Claus, or the Flying Spaghetti Monster. Every person, except little kids who have been lied to by their parents, knows that these three imaginary creatures do not exist. The same cannot be said about God. There is more than ample evidence, as I have shown in these pages for the existence of a divine being Who is outside of our created world. Have you ever heard of a grownup having a Santa Claus moment in which his mind was wonderfully opened and he began to believe in Saint Nick? Yet, there are hundreds of millions of men and women across the globe who profess to having had a God moment.

Tremendous amounts of time and energy have been invested in creating an atheistic replacement for God; such as secular humanism, naturalistic evolution, communism, materialism, or other life philosophies that will help soften the blow of putting God out of our lives. These placebos for God have not satisfied the hearts and longings of men to know their Creator.

C. S. Lewis once remarked that God is not the sort of thing one can be *moderately* interested in. After all, if God does *not* exist, there's no reason to be interested in God at all. On the other hand, if God *does* exist, then this is of paramount interest and importance to our lives, and our ultimate concern ought to be how to be properly related to this Being upon whom we depend moment by moment for our very existence. If God does not exist, why do the non-believers fuss about what other people believe, and why have they been so unsuccessful at eradicating God from humanity's belief systems?

Death –

It may sound strange, but the fact that death exists is one of the strongest proofs for the existence of God. Everything in our world has a beginning and an end as well as a purpose and this tells us that God is a Person full of mercy.

If everything continued on as it is without change forever, then that would be a good reason for saying that there is no God. God is our hope of rescue from the vicious, endless cycle of evil that exists within our nature and in the world. Genesis tells us that he placed a barrier around the Tree of Life so that man could not eat from it and live forever in his present corrupt state.

Though few people look on death as an evidence of the mercy of God, the truth is that we would be trapped forever like this if there were no God.

Man is more than just a physical, biological body. We also have a soul, a personality, a something that is unique in creation, and that tells us that life will continue after our physical death. Though we all go to the grave, that is not the end of our existence.

I ask you to seriously consider the possibility of God as the Person responsible for creating our world and all of the life that is in it, and also the One who offers us hope for deliverance.

52

Our concept of God

Consider what you owe to His immutability. Though you have changed a thousand times, He has not changed once.
Charles Spurgeon

Who is God and what is he like? When we hear the word or name "God," different people are thinking of him in different ways. Every human being has an image/concept of God that is woefully inadequate. What we think of as God is a very incomplete and often distorted view of who he really is.

The apostle Paul said: "*Now we see things imperfectly, like puzzling reflections in a mirror, but then we will see everything with perfect clarity. All that I know now is partial and incomplete, but then I will know everything completely, just as God now knows me completely* (I Cor 13:12 NLT). Even in the Godliest people, their comprehension of him is terribly incomplete.

It is worth saying that our understanding of God does not change who he is in the least. God is constant; unchangeable in his nature and his character. Whether a human being understands or doesn't understand who he is does not make him any different.

Just because your GPS directs you to a wrong address doesn't mean that the real place has moved. What we believe about something doesn't change what it is. We often don't understand the truth about God and therefore we don't really know how to relate to him. Throughout the world, there are as many different concepts of who God is as there are people. This does not make God any less real. It just reveals the level of ignorance of who he really is.

Where does our concept of God come from? In its most basic origin, it comes from the knowledge of who he is based on the created world in which we live.

Paul wrote in Romans 1:18-20: *But God shows His anger from heaven against all sinful, wicked people who suppress the truth by their wickedness. They know the truth about God because He has made it obvious to them. For ever since the world was created, people have seen the earth and sky. Through everything God made, they can clearly see His invisible qualities—His eternal power and divine nature. So, they have no excuse for not knowing God* (NLT).

Because of our rebellious nature, we often do not acknowledge God's authority over us as Creator.

I remember giving a talk to a group of teenagers in Paraguay a long time ago about our concept of God. To illustrate, I had taken about a dozen glass containers of different sizes and shapes, and placed action figures and Barbie dolls in them. Each one was different.

I told the kids that we often keep our image of God inside a bottle or a box, and that every now and then it is necessary to break the bottle in order to let our image of him grow. As I said that, I threw a bottle with an action figure into an aluminum bucket on the tile floor. Everyone was awake, at least for a little while, as the glass shattered.

I told them that unless we can periodically allow our concept of God to grow, it would always pale in comparison with reality. God himself is always so much greater and better than anything we have imagined so far.

53.

The God I believe in

God, the LORD, created the heavens and stretched them out. He created the earth and everything in it. He gives breath to everyone, life to everyone who walks the earth.
Isaiah 42:5

There is no other God—there never has been, and there never will be. God, in Isaiah 43:10b

Based on the world I see around me and the scientific evidence discovered by others, I know that God must be a supernatural Being who was not created like this world was, and who is far superior to it. He must be a Being who is mighty in his power, extremely wise in his creativity, and very interested in his creation. Not only did he make this complex and fascinating world, but he also continues to maintain it every day and every hour.

There is no such thing as "Mother Nature" dictating what happens in the world, but it is God himself who is providing and sustaining all aspects of his creation.

The Scriptures tell us that "*no sparrow can fall to the ground without his being present.*" He abundantly provides food for the birds of the air and the fish of the seas. He regulates the temperature, and balances the composition of the atmosphere.

Many times, he causes or allows seemingly tragic things to happen. Part of the reason is due to the evil that entered into his creation when Satan and humans sinned and perverted what was originally good. Part of it is due to his desire to see humankind reconciled with himself and therefore saved from the destruction that is going to come.

God allows suffering because suffering makes us look heavenward and get our eyes off of just what is in front of us in the physical world. A man can see God much better when he is on his back than when he is healthy and prosperous and only thinking about himself and his own desires.

God truly loves all of his creation, but especially man who was made in his image. "In his image" means that we were made similar to God in several ways. It means that we have intellect that enables us to reason, we experience emotions like he does, and we have a free will to make choices like God does.

We have the ability to have a personal relationship with the God who made us for that purpose. None of the other creatures, except angels were made like that. However, even the angels were not given the privilege of becoming God's children.

God proved his love for humans when he sent his only eternal Son, the Word, to Earth to become a man in order to rescue man from eternal corruption that would send him away from the presence and the goodness of God. That Son became Jesus when he was born of the virgin Mary in the little town of Bethlehem.

After living a life of 33 years of perfect obedience to his Father, He gave his life on the cross to die in payment for man's sins. Romans 6:23 says that the penalty for sin is death, and Jesus died both physically and spiritually two thousand years ago (his relationship to his Father was severed on the cross at three in the afternoon before he cried out "Father, why have you forsaken me?").

In so doing, he made it possible for us, who were condemned to eternal, solitary confinement in darkness away from our Creator, to instead receive what the Son deserved – eternal glory and blessing in the presence of God forever. The resurrection of Jesus is proof that the plan worked. If the account for sin had not been paid in full, Jesus would still be in the grave. But, fortunately for those who truly believe and accept God's offer of deliverance, it is paid, and the resurrection is the receipt guaranteeing our future glory.

The evidence of God's existence is all around us. We need only look at the stars, or through a microscope, or watch a flower grow, or examine a blade of grass in order to see that there is a designer behind the design.

God could have made the world black and white, or made us blind or deaf, but he gave us all the faculties to see, hear, taste, feel and smell so that we can examine his wonderful creation using the amazing minds that he gave us. To ignore the obvious is not science or reason – it is foolishness.

God is patient with us because he loves us. He is working and hoping for us to acknowledge him as our Creator and Sustainer. He is delighted whenever one more of his very special creatures admits that he is on the wrong path and turns and receives Christ's redeeming forgiveness. There is great rejoicing in Heaven for each wayward person who finds the path to life---which is simple faith in God's Son Jesus.

No one can force a person with freedom of choice to believe in something he doesn't want to believe in. Faith comes as an act of a person's will after examining the evidence for some truth. If you refuse to believe in something, no team of horses will convince you otherwise. No amount of scientific evidence will pry your mind loose from prejudice unless you decide to open your mind to consider a different option.

Minds are like parachutes, they only function when open.
Thomas Dewar

All human beings function on faith. Today when you sit down to eat you have faith that your food is not poisoned. When you get into your car to go somewhere, you have faith that you are not going to have an accident. When you marry someone, you have faith that that person is going to be faithful and will be a blessing to you. When you sit on a chair, you have faith that it is not going to collapse and let you fall.

Every scientist, and non-scientist, operates on faith. The question for a person seeking truth about our origins is *"What is the most likely explanation for why I am here?*

Does it take more faith to believe in a Creator, or to believe that impossibly good luck repeated itself millions and millions of times until I was finally born as a result. Which makes more sense and requires less faith?

My faith has found a resting place where the evidence is overwhelming; where the questions of origins are answered, and where I have hope for a promising ending to the story.

About the Author

Myron Loss is a retired missionary who served thirty-five years in Bolivia, Paraguay, and Uruguay with Andes Evangelical Mission and SIM. Prior to that, he was a pilot in the USAF and served for one year in Vietnam.

After marriage for ten months to Janet Beaston, they were involved in a freak auto accident caused by three drunken men who intentionally pushed their car off of the interstate highway near Kansas City, Missouri. Janet died the next day. Myron realized that life is short and eternity is forever and understood that he should invest his life in the things which do not perish or fade away.

His ministries have been varied, including evangelism, discipleship, missionary aviation, Marriage Encounters and marriage counseling, church planting, starting a radio station, construction, candidate recruitment, literature distribution, hospitality, teaching, writing, and administration.

He has an industrial engineering degree from Penn State University, an aeronautical rating from the USAF, a diploma in Advanced Bible Studies from Moody Bible Institute, a Master of Arts in Cross-Cultural Communication from Columbia Graduate School of Bible and Missions, and a Master of Divinity degree from Columbia Biblical Seminary.

He has authored seven other books: *Culture Shock: Dealing with Stress in Cross-Cultural Living* (1983), *The Reward of the Righteous* (1994), *The World Is Upside Down* (2015), *Letters from Heaven* (2015), *God Is with Us* (2015), *Why Many Scientists Believe in God* (2017), *Jesus, my Substitute* (2018), and *Can you Lose Your Salvation*? (2018).

He presently resides in Ocala, Florida, with his wife, Alice. They are members of Oakcrest Baptist Church, and they lead a Hispanic congregation.

